Children of God
Children of Earth

James R. Curry

authorHOUSE

AuthorHouse™
1663 Liberty Drive, Suite 200
Bloomington, IN 47403
www.authorhouse.com
Phone: 1-800-839-8640

© 2008 James R. Curry. All rights reserved.

No part of this book may be reproduced, stored in a retrieval system, or transmitted by any means without the written permission of the author.

First published by AuthorHouse 12/15/2008

ISBN: 978-1-4389-1846-4 (sc)
ISBN: 978-1-4389-1847-1 (hc)

Printed in the United States of America
Bloomington, Indiana

This book is printed on acid-free paper.

To Lorilee, Moss, and Kate

With my love

Contents

Acknowledgements	ix
Introduction	xi
Science	1
1. The Culture of Science	3
2. Science is Human	16
Evolution and the Death of Natural Theology	21
3. The Age of Exploration	23
4. What Hath Darwin Wrought?	39
5. Evolution	51
6. Common Descent	62
7. Why Males?	74
8. Genesis	88
Humanity Unveiled	101
9. Hands	103
10. Handy Man	115
11. Wise One	124
12. Language of the Hands	133
13. Cain and Abel	142
Science Purloined	147
14. Three Generations of Imbecils	149
15. Barefoot Scientist	159
16. Creationism and Intelligent Design	163
17. Return To Mysticism	173
18. Life Happened	186
Environmental Values	197
19. Rapa Nui	199
20. Then Came Christianity	214
Religion	221
21. A Living Faith	223
22. Love Is Eternal	227
23. The God Within	237
24. Spreading the Gospel	259
25. Demons	274
26. Do Unto Others	279
Coda	297
Bibliography	299

Acknowledgements

To all who helped, please accept my heartfelt gratitude.

Franklin College: For much support, both financial and otherwise.
William Pohley: Thanks Bill for always being there when I needed technical help.
William B. Martin: Bill - Your insightful critique of the manuscript was invaluable in keeping me on track. You made a difference.
Richard Curry: Thanks brother for editing the manuscript and for giving encouragement when it was badly needed.
Clifford Cain: Thank you my friend for editing and critiquing the work early on.
James Moseley and David Brailow: Many thanks for granting a sabbatical year to write and dream. I learned things about myself I would otherwise have never known.
Joel Allegretti, Lois Anne Davison, Amanda Stine, and Eve Merrick-Williams: Thank you for permitting me to use your beautiful poetry. It enlivens and accents the text beautifully.
Amanda Boyers: Mandy – You are the best. Life is so much easier when you are around.
William O. Pruitt, Jr.: Thank you for allowing me to recount your story of the Moose People's Winter Festival.
Steve Browder, Alice Heikens, and Sam Rhodes: Thank you for your encouragement and support through the years.
NASA: Thanks for permission to use the stunning cover photo.
Mom and Dad: Thanks for everything! Not a day goes by when I don't think about you and miss you.

Introduction

Aristotle pointed out that it is in our nature to want to know.

There are certain questions we all ask ourselves from time to time:
Why am I here?
Now that I am here, how did I get here?
What is expected of me?
How will it end?
Beyond these very personal questions are those about the larger nature of the universe: Is the universe value oriented? Does it exist for a reason? Does it have a purpose? Is it regulated by natural "law" or by God? Is it chaotic and uncaring – cold, mechanistic, into which our species blundered by chance like a small dog caught in the middle of a busy Los Angeles freeway? Or, does it have immaculate order, created for us by a loving God who knows and cares for each of us as individuals? What is fact and what is wishful thinking?

Perhaps a question larger than any of those above is: How can I answer these questions? Aristotle pointed out that it is human nature to search for answers to life's problems and questions, but not everyone agrees on how best to know. One way of knowing is through science. Many would argue that science is the most powerful way of knowing ever invented by our species. They point to the extraordinary technological prowess garnered from science - from DNA technology to the exploration of space - as proof of the validity of their point of view. Others point to religion as a revealed way of knowing and maintain that it is the ultimate source of knowledge about ourselves and the universe around us.

If we examine the four questions posed above it quickly becomes apparent that science can attempt to answer only parts of the first and last. Evolutionary theory can explain our origins to the satisfaction of many, and death is a part of the natural order. But, is there more even to these two questions that science cannot answer? To many people the answer is yes. They would link all of the questions together, believing that how we got here is related to why we are here, what is expected of us while we are here, and how our existence will end. In their minds, only religion can answer all of these important questions.

To many people the two approaches to knowing are mutually exclusive. Scientists like Richard Dawkins, E. O. Wilson and others see religion as an antiquated crutch that is no longer of any value in the modern world. To them the universe is a cold and uncaring place where natural "law" reigns. On the other hand, the Far Right in religious circles in the United States interprets the Bible to say that God created and maintains the universe in a very direct way. They believe that the universe was created for mankind and is a place of order and meaning. Most people, scientists and theologians alike, occupy some position in the middle ground between these two extremes.

To my mind as a biologist, arguments about the existence and nature of God can never be definitively answered because the supernatural is, by definition, beyond our ability to sense or comprehend. The important questions therefore revolve around the historical and current value of religion to our lives and to the life of society at large. With regard to religion we might ask:

How and why did people begin to believe in the supernatural?

Of what social and personal value are the incredibly complicated and diverse religious rites and beliefs of people through the ages?

On the whole, has the extraordinary time, effort and resources devoted to religious practice benefited humankind or would all this effort have been more productive if spent in other ways, such as in the pursuit of science or in the development of more equitable and cooperative social systems?

How does one understand the violent and bloody conflicts in religious history? Why have people been so quick to slaughter each other because of differing religious beliefs and practices?

Can the writings of people who lived and died thousands of years ago help us to solve the problems facing us today?

With regard to science we might ask:

Has science made us more secure or less secure in our daily lives?

Is a scientific interpretation of the universe and its workings more satisfying and relevant to us as individuals and as a society than an appeal to supernatural explanations for natural phenomena?

Will science be able to address the accelerating decline in the life support systems of our planet, most of which have resulted from scientific and technological "advances?"

Have science and technology sown the seeds of our own demise as a species, or can we come together to build a world that is free of conflict, one in which we will husband the human and natural resources of the planet in a way that is sustainable into the indefinite future?

The problems which beset the world have never really changed: they are as old as civilization itself. What make these problems so much more dangerous today is their scale, as well as the weapons now available for destroying each other. Science has provided us with the means to alleviate suffering from poverty, disease, and hunger, but much of the world's people are still plagued by these age-old problems. Religion has provided us with value systems to guide our behavior and promote personal and social wellness, but that effort has also been largely a disappointment. Some of the most severe threats today come from religious groups whose efforts are directed at forcing others to their way of thinking and acting.

Our species and those pre-human species which preceded us survived for millions of years because they were able to live cooperatively and make biological and cultural adjustments to changing environmental conditions. Whether we will do as well, or whether our failure to adapt to the conditions of a modern world will destroy us, remains to be seen.

This book explores the questions posed above and seeks answers to them. The text of the book is divided into four sections, each of which explores a different aspect of our biological and spiritual heritage and how these relate to the problems of the modern world.

The Culture of Science: Examines in two chapters the history, methods, benefits and limitations of science. Animism, natural theology, and the modern scientific method are discussed.

Evolution and the Death of Natural Theology: Charles Darwin the man, his theory, and his impact on society are profiled in two chapters followed by four chapters which examine modern additions to his theory.

Humanity Unveiled: Five chapters discuss the biological heritage and nature of humanity which are responsible for our uniqueness as a species and which have led to our ability to discern a spirit world beyond our senses.

Science Purloined: Investigates in five chapters some of the past and present abuses of science which have been used to further religious, political, and/or economic agendas.

Environmental Values: Two chapters look at the current state of the world's environments and the problems which are likely to be faced by future generations if science and religion are unable to change the way we relate to the natural world.

Religion: Discusses in six chapters the role of religion to people as individuals and to society at large and suggests that most religions today are failing to adequately address the problems of the modern world and in some instances are responsible for those problems.

Nothing less than a revolution in the values which guide our behavior with respect to each other and to the natural world will enable us to come to grips with our problems and rise above them. This book offers hope that through religion *and* science working together we can do so.

Science

Science is the most powerful way of interpreting the natural world ever devised. First introduced by the ancient Greeks, it flowered in Europe following the Middle Ages and has radically changed the way we live and think about ourselves and the universe. Science and its offspring technology have been critical in our efforts to survive and prosper in a world that is often difficult to understand and impossible to control. The following two chapters include a brief history of science, highlighting major advances in method and knowledge.

CHAPTER 1

The Culture of Science

For man by the fall fell at the same time from his state of innocence and from his dominion over creation. Both of these losses, however, can even in this life be in some part repaired: the former by religion and faith the latter by arts and sciences.

- Francis Bacon

Every living organism must be able to interpret its world. Any failure in this ability means certain death for individuals and eventual extinction for its species. Even the simplest forms of life like bacteria and other single-celled organisms can respond appropriately to positive or negative stimuli in their environment. Because of this ability, they are able to feed themselves, escape predation, find mates, and thus survive in space and time. Plants, which may seem to be completely unresponsive to their environment, are able to send roots toward water, bend toward the light, flood their leaves with chemicals in response to insect attacks, adjust their size and shape during growth to fit the environmental conditions, reproduce at the most appropriate time of the year, and lose or grow leaves in response to seasonal changes.

As humans, we have from the very beginning been able to observe and interpret the natural world to an extent not possible by other species. The survival of early man was dependent upon his ability to search out plants that would feed but not poison him; hunt animal prey successfully; prepare for the approach of winter or other hard times; know when to move and when to stay; find appropriate materials for the making of tools, clothing, weapons, and shelters; and escape predation by other animals. Our survival has always depended directly upon the ability to do these things well, and to pass that knowledge on to succeeding generations. As our ancient ancestor's knowledge of the world increased, so did their success. They multiplied their numbers and began to migrate out of Africa to other parts of the earth. Equally important to the knowledge about the world around them, was the knowledge of how to observe and interpret that world. Much of their success was due to their increasing knowledge of how best to study the world, and how best to communicate this knowledge to each other and to their offspring.

The two most important discoveries in human history were a direct result of this growing ability to observe and interpret the natural world: agriculture (including the domestication of animals), and science. Both are relatively recent events in human history. Nobody knows, or is likely to ever know, how the knowledge of propagating plants was first learned, but most certainly it was the culmination of a long history of closely observing plants to determine which were good to eat, which had medicinal properties, which produced fruit and in what season, and which grew where. Accidental learning has always been important to the growth of human knowledge, and perhaps the knowledge that the seeds of favorable plants could be harvested and grown was obtained in this way. Seeds carelessly discarded around the campsite began to grow, and by trial and error, those which grew best and were most productive were selected and planted. Thus, perhaps, began the technology of agriculture, and it completely revolutionized human culture.

With agriculture, which appeared about 10,000 years ago, the hunting-gathering way of life was gradually abandoned by most groups of people. More or less permanent dwellings were constructed and settlements began to develop. Tools for working the ground were devised, vessels for storing food were developed, and humans prospered as they never had before. It is estimated that prior to the development of agriculture the total human population of the world was not more than 5 million people. Agriculture provided far more food than was ever possible with a hunting-gathering existence and the population began to increase dramatically.

The technology of agriculture was so successful, that within a few thousand years writing, art, and architecture flourished in a way that was impossible in a hunting-gathering culture. Huge numbers of people were freed from the toil of producing food to build structures like the pyramids of Egypt and Central America and the other great wonders of the ancient world. Feeding twenty thousand people once or twice a day for decades while such gigantic projects were completed was an enormous task but the ability to grow crops, to store food, and to domesticate animals for labor made it all possible.

Agriculture made the human species wealthy with the only wealth that really counts: food. As the technology which spurred such wealth was being developed, so was the knowledge of how to observe and study nature in order to further improve the human condition. At the same time, writing provided a better method of communicating and passing on what was being learned than had the oral traditions of the past. All of this culminated in the sixth century B.C. with the invention of science, which through the centuries has developed into an extraordinarily powerful tool for observing and interpreting the natural world around us. Science is a careful, methodical, organized way to approach the study of nature. It has given us a remarkable power to investigate, interpret, and describe the natural world and to use this knowledge to create technologies that make us more successful at surviving. As in the past, the knowledge of how to study the world continues to increase along with the knowledge obtained about the world. Science, the method, and knowledge, the result, continue to grow exponentially.

Animism – Precursor to Science

Ancient people often resorted to supernatural explanations for natural phenomena. This was not because they were stupid but only because they lacked the knowledge to understand the underlying causes of natural occurrences. They made no distinction between living and nonliving things, believing that mountains, rivers, celestial bodies, rocks and all other inanimate objects have souls. Anthropologists call this animism. Animism was universal among ancient peoples and persists today in many different forms in some parts of the world. Many anthropologists consider animism to be the original human religion, and most religions today are tinged to a greater or lesser degree with animistic beliefs and rituals.

Holy men or women, sacred items, sacred spaces for worship, rites to appease or otherwise influence the spirits, and ancestor worship are all characteristic of animistic societies. In studying animism it becomes clear that ancient humans felt very much out of control in a natural world they understood very poorly. Their efforts to explain and gain a measure of control over nature through animism seems extraordinarily naive to us today in light of what 2600 years of scientific study has taught us, but it was all the ancient people had to relieve the fear and anxiety of facing a threatening and capricious world.

Animism is apparently a very old practice in human societies. Doll-like fertility figures representing mother earth have been found that are more than 25,000 years old. The belief in a fertility goddess who could produce or withhold babies was common in ancient cultures.

After the introduction of agriculture, the increased food supply provided greater security, but people had to work harder and longer than hunters and gatherers. One of the common myths today is that people who live in hunting/gathering societies are poorly fed and live on the edge of starvation. Nothing could be farther from the truth. Such people are not only well fed and healthy, but most are able to provide for their needs by working only a few hours each day. People living in agricultural societies where each family must work its small farm in order to survive work almost incessantly and are often poorly fed due to the reliance on single food items that do not provide complete nutrition. A steady diet of rice, corn, potatoes, cassava, or any of the other common food crops does not supply enough protein for growth and development, and as a result children are often stunted in their physical and mental growth. Working a hardscrabble farm with the crudest of tools is a sure path to a broken body and an early death.

Once populations grew as a result of the conversion to agriculture, it was no longer possible to return to a hunting/gathering way of life. If crops failed, an entire year's worth of food was lost and starvation quickly followed. Spring was a very special time of year even before the birth of agriculture, for it was a time when life was renewed. It became even more important after the advent of agriculture because this was a time of planting, and life itself was dependent upon the successful growth and harvesting of the crops. Fertility rites, already ancient, took on a new meaning and importance. These were often held in the fields to encourage the growth of crops as well as the fertility of people.

Modern farmers can sympathize with these ancient people, for even today when an entire year's income depends upon a good crop, there is little a farmer can do to control the natural forces that will determine his success or failure. Some of the rites surrounding the advent of spring survive today. The egg has been a symbol of fertility since the dawn of man. Colored eggs were given as gifts during ancient spring rituals in ancient Egypt and Persia, a practice that survives today at Easter. May Day celebrations such as dancing around the maypole are also ancient fertility rites of spring.

Our ancient ancestors were not stupid savages. Their belief that everything in the natural world, whether animate or inanimate, had a spirit that controlled how it behaved helped them to accept, and in a sense control, that which to them was uncontrollable. If rain did not fall on their crops, making sacrifices to the rain spirit, if nothing else, helped them to feel that they were "doing something". If the sun, the giver of life, disappeared in an eclipse, rites designed to make supplication to the sun god gave them some sense of security. Knowing nothing about the causes of disease, they thought that the gods were angry, and devised various rituals to appease them so that the sick person could recover. When sick people did recover, this tended to reinforce the belief in gods as the cause of the disease

We should not forget that in a sense the ancients were far more attuned to the natural world, and understood it better, than most people today. As the world becomes more and more urbanized, fewer and fewer people come into direct contact with nature. Viewing the stars or feeling a fresh wind while standing in a city is nearly impossible. Nature exists in its most artificial form in our cities, and people who spend their lives there know less about nature than the most ignorant of our ancient ancestors who used stone tools, wore animal skins, and tried to control natural forces by making offerings.

Animism didn't portray the world as duality – natural and spiritual. One world view combined the knowledge of trees, winds, animals, mountains, and rivers with a spiritual world full of hope, fear, and imagination. Modern people are prone to divide the world into two different, often competing entities: the natural, and supernatural. In many religions, the only thing that connects the two is man himself. Man alone is seen as having a body and a soul, but the body had a beginning and will have an end, while the soul or spirit is without beginning or end. Because man is the only part of the creation that will transcend the natural, nature is often relegated to a temporary and disposable position in thinking. What is more, since we alone in the entire universe can transcend nature, it must have been created for us.

Early humans believed that everything had a dual nature - a material substance, and a supernatural spirit, but they did not separate the two in the same way that people do today. In a sense, animists integrated nature with the supernatural: each "thing" in the world contained elements of both, and the two interacted in a way that explained its substance and behavior. We tend to think that early man lived only for the moment or the immediate future, but his worshiping of ancestors, which was an integral part of animism, helped him to see beyond the moment and combine the past with the present and future. The spirits of the dead lived on, not in heaven or some other place removed from the earth, but with the living. These spirits watched over the living but also demanded attention and had to be appeased the same as other spirits.

To the animist, all life was sacred; indeed, all of nature was sacred. Prior to cutting down a tree a prayer was said asking the forgiveness of the tree's spirit. When an animal was killed, a prayer was said asking forgiveness and explaining that the hunter did not want to kill but was forced to do so to feed his family. With the general loss of animism came the loss of much of our respect for life other than human life. It is a loss to be regretted.

How successful was animism? The fear, imagination, passion, self-identity, and stubborn desire for control that were inherent in the practice of animism led to the early development of art, architecture, religion, agriculture, metallurgy, writing, technology and science. These are the hallmarks of our humanity. But animism did not provide the security and freedom from fear that early peoples wanted and needed. The gods were capricious and unpredictable. They could even be intractable in the face of sincere supplication and appeasement. The god Pele, for example, god of fire to the Hawaiian Islanders, was an extremely important god because she lived in the volcanoes that produced land, an important commodity to people living in the middle of an ocean. But, she was uncontrollable, and could be as destructive as she was helpful. There was no way to control or even understand her. The anaconda, worshiped as a god by peoples of the Amazon basin, would frequently carry off children as they bathed or played in the streams, and despite all the supplications of grieving parents, would not return them. Nonetheless, the snake was an important element in their society, and they needed to understand it, and in understanding, gain a measure of control.

The solution to the problems posed by animism was slow in coming and incomplete in its solution, but gradually the ancient peoples came to realize that there is a consistency in nature that seemed to defy their gods. They noted that the sun always rises in the east and sets in the west; the annual rains come the same time each year; the stars and moon move through the heavens in a predictable way and can be relied upon to tell both seasonal and daily time; the plants and animals go through their life cycles of reproduction and growth in a patterned and predictable manner. The ancients had always been close to nature, but their knowledge of it was not organized. Careful observation and experimentation organized by logical thinking are the methods of science, and people were moving in this direction, but they could not really develop a scientific approach to nature until agriculture, writing, and mathematics were invented. Nor could they develop the great modern religions until a scientific approach to the study of the universe made their animistic gods less effective. Only then could the powerful concept of one God develop. By separating the natural from the supernatural, and studying each independently through science and religion, a unifying concept of the universe was born which is still evolving today and which works much better than animism in alleviating fear (or increasing security) and in solving the everyday problems of life.

Ancient Greece and the Origin of Science

One of the milestones in the development of human culture, second only to the discovery of agriculture, was the invention of science as a method for observing and explaining natural phenomena. The breakthrough is thought to have occurred in the sixth century, B.C. in Greece and is referred to as the "Greek Miracle." The slow transition from animism to Greek science required a revolution in the way humans viewed the world.

Animism gave no unified concept of the world. The spirits which were supposed to control and pervade nature in animistic thought were individual and independent of each other. The concept provided no coherent world order through which nature could be viewed and studied. Animism also did not provide a moral code; it was (and is) without ethical thoughts and motives. The development of organized religion apparently proceeded from this latter need. Thus, the two most important aspects of our modern culture, religion and science, both sprang from the demands of a rapidly developing agricultural society, demands which could no longer be accommodated by animistic thinking.

Although science began in Greece, it should not be forgotten that there were elements of a slow transition away from animism in the civilizations of Mesopotamia and ancient Egypt. The ancient Egyptians deified nature much as had earlier cultures. Ra, the supreme sun-god, was considered to be the father of all creation and was usually depicted as a falcon or lion. Horus, the sky-god was ruler of the day and was also depicted as a falcon. Hathor, depicted as a cow, was the goddess of the sky, of women, and of fertility and love. Nut was the mother of the sun, moon, and heavenly bodies. Sekmet, was the mistress of war and sickness. It was common for the Egyptian gods to be symbolized as animals or to have the heads of animals, and many animals, both wild and tame, were considered to be sacred because they were favored by the gods. Persons killing sacred animals were punished, sometimes severely. According to the Greek historian Herodotus, the penalty for killing a falcon or ibis was death. Mummification of animals was common, and tens of thousands have been found by modern archaeologists. As in earlier animistic cultures, the human race was not considered by Egyptians to be superior to the animal world. Both were created by gods to share the earth.

The deification of animals and of natural objects like the sun and moon were very reflective of the animism of earlier cultures, and yet the Egyptians were beginning the transition to interpreting the world in natural terms. This can be seen in their medicine. They developed considerable expertise in some forms of medicine and surgery, but continued to use magic charms and incantations in treating disease. Very often a priest was present whenever a physician treated a patient, and magic charms and incantations were administered along with more practical treatments. A case in point is the description for the treatment of burn in the Georg Ebers papyrus found in Egypt in the 1970's:

"Milk of a woman who has borne a male child, gum, and ram's hair. While administering this mixture say: Thy son Horus is burnt in the desert. Is there any water there? There is not water. I have water in my mouth and a Nile between my thighs. I have come to extinguish the fire."

In the case of wounds, skin infections, or other ailments in which the cause was obvious, treatment was practical and rational. According to the Ebers Papyrus, a wide variety of medicines were used, many of them herbal, for the treatment of ailments. Belladonna was used as a pain killer; aloe vera for burns, ulcers, skin diseases, and allergies; honey in dressing wounds; tamarind as a laxative; and so forth. Many of these were undoubtedly effective: honey, for example, is an effective antibacterial agent and will help to prevent wounds from becoming infected. Internal diseases were not well understood, and were attributed to evil gods or divine punishment. The prior development of writing enabled the Egyptians to accurately record the symptoms and cures for disease, making it possible for them to progressively improve their medical practices.

The Egyptians developed a rudimentary knowledge of anatomy and physiology, probably as a result of their practice of embalming. They knew the position of the heart in the chest and understood the relationship between the heart and pulse as indicated by statements from the Edwin Smith papyrus as translated by J.H. Breasted. The papyrus has been dated to 1600 B.C. and it contains 48 systematically described case histories ranging in severity from minor wounds and illnesses to gaping wounds to the skull which exposed the brain. The following statement is from a case history describing treatment for a head wound:

"There are canals (or vessels) in it (the heart) to every member. Now if the priest of Sekhmet or any physician lay his hands (or) his fingers upon the head, upon the back of the head upon the two hands, upon the pulse.... he measures the heart, because its vessels are in the back of the head and in the pulse; and because its pulsation is in every vessel of every member."

This rough translation makes it clear that the Egyptians understood that blood vessels connected the heart to all organs of the body, and that it pumped blood through them.

Describing treatment for a much more serious wound (A gaping wound in the head, smashing the skull), the writer of the papyrus advises: "Thou should say regarding him - An ailment not to be treated." In other words, the outcome of the injury is going to be fatal and the physician/sorcerer should not attempt to treat the wound. This is curious in that no supplication is to be made to the god Sekhmet on behalf of the injured person. This comes very close to being a naturalistic approach to medicine and the human body and may represent the early beginnings of a movement away from medical treatment by appealing to the gods and exorcizing demons by chanting incantations. Still, the ancient Egyptians never were able to come to the recognition that all disease, injury, and other ailments have purely natural causes and should be treated as such. Two thousand years would pass after the scribing of the Edwin Smith papyrus before this revolution in medicine would occur, and it would be in ancient Greece, not Egypt, that the breakthrough was made.

The "Greek Miracle" was the development of an orderly and rational structure for the study of natural phenomena without resorting to mythology or the gods. The sudden emergence in Greece of a natural philosophy in the 6th century B.C. was the beginning of a scientific method which relied on reason and logic for the study of natural phenomena. Plato and Aristotle were two of the most influential people in the history of science, but many others also contributed in astronomy, mathematics, medicine, geography, and other fields. Anaxagoras (499-428 B.C.) was the first to propose that the moon shines by reflected light from the sun, which he referred to as a "red-hot stone". He also correctly explained the cause of eclipses of the sun and moon. This is a far cry from the cosmogony of earlier civilizations like the Mesopotamians and Egyptians which believed that the sun was a god who had to be worshipped and appeased.

Eratosthenes was a Greek mathematician and astronomer who measured the circumference of the Earth with extraordinary accuracy and who created a catalogue of 675 stars. Heraclides in the 4th century B.C. proposed that the apparent rotation of the heavens was due to the rotation of the Earth on its axis rather than the passage of the stars around the Earth. Aristarchus (270B.C.) proposed a heliocentric model of the solar system; that is, that the sun, not the Earth, is at the center and the planets revolve about it. It is interesting to note that during the Middle Ages more than 1500 years later European scientists still believed that the sun revolved about the Earth. Greek science was completely lost to Europe during the Middle Ages. In the 16th and 17th centuries a heliocentric theory was revived by Copernicus, Galileo, and Kepler, but it was extremely controversial and Galileo spent the latter years of his life under house arrest by the church because he refused to recant the theory.

Euclid (323-283 B.C.), a Greek mathematician, is considered to be the "father of geometry" and one of the most influential mathematicians of all time. His best known work, *Elements*, founded the axiomatic method of mathematics and set a standard for rigor that continues to characterize mathematics. Euclidian geometry has been influential in every branch of science. The *Elements* continue to be studied today and it is considered by many to be one of the most influential works in the history of science.

Euclid – Painting by Raphael

The Greeks also abandoned the belief that disease was caused by demons or angry gods. Hippocrates treatise, *On the Sacred Disease* (epilepsy), written in 400 B.C., begins with these words:

It is thus with regard to the disease called Sacred: it appears to me to be no wise more divine nor more sacred than other diseases, but has a natural cause from the originates like other affections. Men regard its nature and cause as divine from ignorance and wonder, because it is not at all like to other diseases. And this notion of its divinity is kept up by their inability to comprehend it, and the simplicity of the mode by which it is cured....

- Translated by Francis Adams

Scientific Revolution

For eight hundred years Arabic would remain the major intellectual and scientific language of the world.
— Kenneth Humphreys.

The scientific revolution which swept over Europe in the seventeenth and eighteenth centuries had its earliest beginnings in the twelfth and thirteenth centuries with the reintroduction of Aristotelian philosophy by Islamic scholars. What Europeans refer to as the "Dark Ages" could better be labeled the Golden Age of Islam. Islamic scholars copied and preserved the writings of Greek philosophers and beginning about the seventh century made important contributions in astronomy, physics, medicine, engineering, and mathematics. During the 9th and 10th centuries, while Europe stagnated culturally under Christian domination, Islamic scholars established the basis for modern science, mathematics, and medicine, and made huge contributions in these areas.

"The scientific method in its modern form arguably developed in early Muslim philosophy, in particular, using experiments to distinguish between competing scientific theories, citation ("isnad"), peer review and open inquiry leading to development of consensus ("ijma via "ijtihad"), and a general belief that knowledge reveals nature honestly."
— Wikipedia.

The scientific revolution in Europe that is associated with Copernicus, Francis Bacon, Galileo, Isaac Newton and other natural philosophers developed from changes in the way Europeans looked at the natural world, largely as a result of Islamic influence. The changes in world view which led to the scientific revolution in Europe were also the result of changes in society as a whole: new lands were being colonized; rebellion against the stifling authority of the Catholic Church was leading to the Protestant Reformation; and suddenly the world seemed larger and less threatening. Although the terms "science" and "scientist" would not come into use until the mid-19th century, many of the hallmarks of modern science were emerging during this period of societal change in Europe. This was also a time of renaissance in art, architecture, literature, and political thought.

The medieval world held the view that knowledge comes only from the intellect and that the senses and experience could not be trusted. The following anecdote attributed to Francis Bacon illustrates this way of thinking:

James R. Curry

In the year of our Lord 1432, there arose a grievous quarrel among the brethren over the number of teeth in the mouth of a horse. For thirteen days the disputation raged without ceasing. All the ancient books and chronicles were fetched out, and wonderful and ponderous erudition such as was never before heard of in this region was made manifest. At the beginning of the fourteenth day, a youthful friar of goodly bearing asked his learned superiors for permission to add a word, and straightway, to the wonderment of the disputants, whose deep wisdom he sore vexed, he beseeched them to unbend in a manner coarse and unheard-of and to look in the open mouth of a horse and find answer to their questionings. At this, their dignity being grievously hurt, they waxed exceeding wroth; and, joining in a mighty uproar, they flew upon him and smote him, hip and thigh, and cast him out forthwith. For, said they, surely Satan hath tempted this bold neophyte to declare unholy and unheard-of ways of finding truth, contrary to all the teachings of the fathers. After many days more of grievous strife, the dove of peace sat on the assembly, and they as one man declaring the problem to be an everlasting mystery because of a grievous dearth of historical and theological evidence thereof, so ordered the same writ down.

Such thinking began to give way as early as 1214 with philosophers like Roger Bacon (1214-1294), who believed that knowledge proceeds from observation and trial and error methods. During the 16th century Copernicus put forth the theory that the sun, not the Earth, was the center of the solar system and that the Earth revolved upon its axis rather than the stars and sun moving around the Earth. This was not accepted during his lifetime and it remained for Galileo to win wide acceptance of the theory. Isaac Newton's work provided a mechanistic view of the universe, one controlled by natural laws rather than by the hand of God. Newton and the other natural philosophers were intensely religious, but the restrictive hand of the church was slowly relaxing its hold in the search for knowledge about the natural world.

It was in the 19th century that modern science was born. The term *scientist* was coined by William Whewell in 1833, and the century produced an extraordinary number of brilliant scientists including Amedo Avogadro, Marie Curie, Charles Darwin, Christian Doppler, Michael Faraday, Alexander von Humboldt, Lord Kelvin, Robert Koch, Gregor Mendel, Alfred Nobel, Louis Pasteur, and Nikola Telsa. Among the most influential of these were Charles Darwin and Louis Pasteur.

Pasteur's phenomenal contributions included a germ theory of disease, the discovery of viruses and immunization, pasteurization to destroy harmful microorganisms in food, an understanding of fermentation, including the discovery that life could exist without oxygen, and the destruction of a widely held belief that life could arise spontaneously from non-living materials. It was through rigorous experimentation that Pasteur developed his germ theory, but more than that, his work brought about a revolution in scientific methodology. He exhorted his disciples: "***Do not put forward anything that you cannot prove by experimentation.***" This has become the ground rule for all scientific investigation since Pasteur's time.

CHAPTER 2

Science is Human

Space, time, causality, and physical substance - all are human constructs which alien species from another world would not understand as we do. Does our knowledge have any intrinsic value? Would it be of value to an alien life form. No. If our species should become extinct, all that we know - mathematics, astronomy, biology, physics, chemistry, music, art, philosophy, religion – would die with us. It is not universal truth, it is totally humanistic, and as such, meaningless to other life forms. Their perception of the universe would be as different from ours as ours is from a jellyfish; perhaps even more so, because we share a common evolutionary heritage with the jellyfish.

Science as a cosmology is based upon the assumption that there is, in Alfred North Whitehead's words, *an Order of Things*, and, in particular, an *Order of Nature*. Furthermore, the assumption is widespread that we are capable of interpreting that order in realistic terms. Both of these assumptions are so ingrained in our culture as to go unchallenged today, but, they are both assumptions which derive from what may well be an unfounded faith in our own capabilities. If either or both of these assumptions are false, then what Whitehead calls our "instinctive conviction" in their validity is very badly misplaced. He goes on to state that our faith that order exists in nature is "impervious to the demand for a consistent rationality."

My memories of early childhood are faded and selective. One memory of my preschool years stands out very vividly, however. I remember thinking that people and objects were active only when I was in their presence. If I walked into the kitchen and my father was sitting at the table with a cup of coffee and my mother was washing dishes at the sink, I pictured them as having been frozen in place before I walked in. The instant I entered, my father would raise his coffee cup to his lips and my mother would rinse a dish. People existed only for me. I guess it is not unusual for small children to be so self-centered.

Apparently I was not completely convinced of my theory, because I clearly remember trying to find a way to prove or disprove it. Peeking around corners did no good because the instant I injected myself into a situation it became animated. I finally realized that my belief could not be proven or disproved. As I grew older I learned that people existed before me, which was even harder to accept than that people had a life apart from my own. Eventually, I gave up on the whole concept and accepted that people exist on-their-own and have done so for a very long time.

What I came to accept as a child is simply that the world does exist apart from me. This is also a basic assumption of science. Without it, science could not exist. And yet, it is no more provable to a scientist than it was to me as a small child. It is an assumption, taken by faith; hence, it is not questioned.

This is not to say that scientists assume that the "real" world exists as we perceive it. Something exists apart from the observer, but in all probability what is perceived is at best a severe abstraction of reality. The objects we perceive in our environment are intuitions of the mind, mental constructs formed from a complex of sensory input and training. These mental images are more than the sum of the individual bits of sensory information relayed to the brain, for it is intuition, (some might say instinct), that transforms the bits and pieces of sensory input into a functional mental entity. The difference between sensory input and the mental image of an object is like the difference between a pile of stone and the Taj Mahal. The brain takes bits and pieces of sensory input and makes them into something meaningful, but it also removes them one step further from the reality of the object being experienced.

Imagine standing in front of a tree and describing in detail what the tree looks like to someone who has been blind since birth. Step by step a mental image of the tree would form in the mind of the blind person from your description. How closely would the mental image of the tree in the blind person's mind resemble your own? We can only guess at the answer. Having never directly experienced colors and shapes the blind person's conception of such things might produce an image of a tree that is quite different from your own. On the other hand, since he has a brain that has evolved over millions of years specifically to build images of color and shape, and since these are mechanisms internal to the brain itself, perhaps the blind person would in fact build a mental image not too different from yours.

The image the blind person conceives is twice removed from the tree itself, having been translated once by your brain, and once again by his brain from your description. Which image, in which person, represents the "real" tree? The answer is both. Each is undoubtedly a severe abstraction of the "real" tree, but each has a reality of its own which is quite removed from the tree itself.

Like a picture puzzle, the tree image would grow piece by piece in the mind of the blind person. The pieces of the puzzle are the words of the describer. Words are mental images themselves. The word dog is not really a dog, but it conjures up the mental image of a dog in our brain. The image of the tree in the brain of the blind person will be shaped by the words of the describer - piece by piece - until a picture emerges. We have all had the experience of having some object we have never seen described to us, and upon actually seeing the object for the first time, immediately recognize it. We have an advantage over a person blind since birth, however, in that we have a lifetime of visual experience behind us and can immediately recognize the description of shapes, colors, and size.

The Scientific Method

I sometimes take students in my undergraduate biology class to a spot on campus where a large downspout runs down from the roof of a three story building. The lower end of the downspout is about a foot above ground, and water shooting out of it during rains has carved a fairly sizeable ditch which continues down a slope from the corner of the building to a sidewalk below. I divide the students into small groups and ask each group to independently examine the ditch and propose a hypothesis about how it formed. I tell them also to defend their hypothesis with evidence, and to propose experiments to test it.

All of the groups immediately look over the situation carefully and come to the conclusion that water coming down the pipe has eroded away the ground and created the ditch as it flowed down toward the sidewalk. Some of the groups will discover small pieces of slate in the ditch and use this as proof that water coming down the pipe from the roof created the ditch (The building has a slate roof). A group will sometimes suggest putting a hose in the pipe at its top to test the effect of water on the soil at the base of the building. Others will suggest returning during a rain storm to collect water at the bottom of the ditch to see if it is muddy with soil. All, however, intuitively know that water created the ditch, and this is undoubtedly based upon their prior experience with moving water and its effect on soil.

After each group explains their conclusions to me, I play the devil's advocate: "I believe that children in the neighborhood have dug out the ditch with their toy shovels and trucks," I say with a straight face. Prove me wrong. The students are then prone to repeat all of their arguments, but of course, they cannot prove me wrong. In fact, if I were to say that I thought little green men from Mars dug the ditch they could not prove me wrong, yet they will not concede that I am correct in my assertions. They are convinced that water coming off the roof dug out the ditch and they will prove it to me if we will visit the site during a rain storm. I counter that the only thing that will prove is that water **can** move soil; it will in no way prove that that is how this particular ditch came to be. My hypothesis that children dug it may not be as intellectually satisfying as their hypothesis, but no amount of testing of any kind can prove theirs or disprove mine.

The point of the exercise is to demonstrate how science studies and interprets natural phenomena. Careful observation, the formation of a hypothesis, a search for evidence that will confirm or disprove the hypothesis, an application of information and knowledge previously gained through experience, and an assumption that there is order in the universe. This last condition is critical to the whole process, but it is still only an assumption. In the case of the ditch, it is assumed that if a test shows that water coming down the pipe today will carry away soil, the same would have been true yesterday and all the days before. It also assumes that, all other things being equal, water coming down the pipe in the past would have had the same force behind it as it does today.

One could go further with this little exercise and by measuring how rapidly soil is carried away during a rainstorm, estimate how long it took to dig the ditch. Once again, no one could prove that the rate of erosion in years prior to the tests was the same as it is during the test.

Now, apply this same procedure to a much larger ditch, say the Grand Canyon in Arizona. A geologist standing on the rim looking down into the canyon might immediately come to the conclusion that it was dug by the forces of erosion - rain, snow melt, tributary streams, and the Colorado River. As proof, he might point to the sediment being carried by the river and to the river's ability during flood to move even huge boulders. He might even measure the depth of the canyon from a platform in space (satellite) and show that it is in fact very slowly getting deeper as the river continues to erode away at its base. None of this, however, will prove in any sense that this is in fact how the canyon came to be; it is merely evidence of what is happening today. If someone chooses to believe that the canyon was dug out by extraterrestrials or a world-wide flood, no scientist can conclusively disprove that hypothesis because no one was present while the canyon was being created.

The geologist might also attempt to estimate the time it took to dig out the canyon by measuring the current rate of erosion and extrapolating back into time. He might date the rocks at various levels in the canyon and use this to estimate its age. In so doing, he would conclude that it took millions of years for the canyon to reach its current depth. If someone chooses to believe that such a thing is impossible because the earth is only six thousand years old, there is no conclusive way to prove him wrong. The problem with explanations based upon suppositions like this is that they do not lend themselves to any kind of examination or experimentation. One must accept such beliefs and suppositions purely on faith, and must hold to them no matter how much scientific evidence exists which point to natural causes. The greatest weakness of this approach to understanding natural phenomena, however, is that they provide no real power of prediction and thus cannot empower us in our efforts to survive and prosper. They are not much different than trying to learn the number of teeth in a horse's mouth through conjecture rather than by counting the teeth.

Through experimentation Benjamin Franklin showed conclusively that lightening was electricity, and having gained this knowledge used it to develop lightening rods which could protect buildings from being burned down by lightning strikes. Fire was an even greater scourge in Franklin's day than it is today. Fire departments were little more than bucket brigades and once a fire started it could sweep through and destroy an entire city. Many of the clergy of Franklin's day were outraged by his invention. They believed that lightening was the wrath of God on a sinful people and that attempting to control it was tempting God, but Franklin's invention empowered people in a way that the clergy's belief could not: it protected them from lightening-caused fires. It is ironic that lightening rods soon appeared on the roofs of most of the churches of the day and that unquestionably many of these buildings were spared destruction by lightening.

These are the methods and benefits of modern science. Science is an invention of the human mind, the most powerful one ever devised for interpreting the natural world. It is the engine that drives the technology which controls our lives. From medicine to space exploration, to microwave ovens, basic scientific research has led the way and will continue to do so as long as our species survives, but there are limits to what science can contribute to our lives and to society at large. Will Durant clearly speaks to these limits in his book, *The Reformation*:

Science gives man ever greater powers but ever less significance; it improves his tools and neglects his purposes; it is silent on ultimate origins, values, and aims; it gives life and history no meaning or worth that is not canceled by death or omnivorous time.

Evolution and the Death of Natural Theology

Animism defined the relationship between people and the natural world in a very direct way: everything had a dual nature – a material substance and a spirit. The spirits that animated mountains, animals and plants also animated humankind. Through rituals and offerings to the spirits animistic peoples attempted to understand and establish a relationship with the natural and supernatural worlds. Since they made no distinction between themselves and the rest of nature, animism at its most elemental level was really an effort by people to understand themselves.

When the ancient Greeks began the process of separating the study of nature from the supernatural a quantum change in peoples' perception of themselves was begun which continues to this day. There were many important steps along the way, and some of these will be explored in future chapters, but one of the most important was the development of evolutionary theory.

Charles Darwin did not initiate the search for an explanation to the origin and development of life on Earth but he set the base upon which modern evolutionary theory is built. His five year voyage around the world resulted in a theory that forever changed the way we look at ourselves and the natural world.

CHAPTER 3

The Age of Exploration

The small ship was rolling violently in the waters off Tierra del Fuego as the young, sandy-haired Englishman clung to the ropes and looked at the rugged, snow covered mountains that came down to the sea's edge. The ship's captain, Robert Fitzroy of the Royal Navy, stood beside him, his face beaming with excitement. They watched together as the ship's crew ferried supplies from ship to shore in the strong currents. This was dangerous work in the strong winds and rough seas. The weather here at the southern tip of South America is among the world's most violent, and it can change in a matter of minutes. Off on the horizon the shapes of large icebergs could be dimly seen through the fog and snow.

The year was 1833, and the young man standing beside Captain Fitzroy was Charles Darwin. Only a few months before Darwin was still a college student at Oxford. He had never been outside of England; indeed, he had never traveled more than a hundred miles from his home. Any excitement he felt at being in this strange and exotic place was clouded by the violent seasickness that had plagued him since leaving England a month earlier.

Nothing, not the weather, the difficulty of transporting supplies ashore, the rough, unpredictable seas, or Darwin's illness could suppress Fitzroy's excitement. He had looked forward to this day for more than three years, and he was determined that nothing was going to ruin it for him. In 1830 while on a hydrographic survey he had induced four young native Yaghans from Tierra del Fuego to return to England with him where they were considered to be little more than half animal. There at his own expense, they were placed under the care of a minister, taught the English language, dressed in the finest styles of the day, and given the names Jemmy Button, Fuegia Basket, York Minister, and Boat Memory. Now Fitzroy was returning three of them to their native land.

Fitzroy was a devout Christian, and his reason for the entire enterprise was to use the three educated natives to spread Christianity and western culture to their people. For this purpose he had also brought along a young missionary, Richard Matthews, who was to be left behind when the ship sailed. The effort, as misplaced and ill advised as it may seem today, was in keeping with the temper of the times. As British influence spread around the globe it was accompanied by increased missionary efforts to save the souls of peoples considered to be savage pagans.

With some difficulty Fitzroy, Darwin, Jemmy Button, York Minister, and Fuegia Basket settled into the ship's boat as it rose and fell in the heavy swell and banged against the side of the ship. The men at the oars strained to row the boat ashore, and eventually they all stepped onto the pebble-strewn beach. Everyone was wet through, and the cold wind and snow flurries bit through their wool clothing.

They were met by a small group of the wildest looking people Darwin had ever seen or imagined. Men, women and children were absolutely naked, although a few of them had a seal skin over the shoulder or a small skin tied around their waist. Snow melted unnoticed on their dark brown skin, and their stiff black hair stood on end. Their face and bodies were painted with red and white chalk, with white circles around their eyes.

Jemmy Button recognized his brother among the group and stepped forward in his formal English suit, top hat and boots, and tried speaking to him in English. When his brother didn't respond he spoke in Spanish. It was an awkward moment as the half wild natives stared at the three returning Fuegians in their finery and wondered what was happening. They had once been friends, but now they didn't even recognize each other.

Eventually they all were led to a small group of huts made of tree branches. These were about four feet high, rounded on top, with a hole in the side just large enough to crawl in on hands and knees. A small fire was kept burning inside, but the huts gave almost no protection from the strong winds and cold. During the next few days small huts were built by the sailors and Matthews. The weather cleared and relations with the Fuegians improved enough that after five days Fitzroy felt comfortable sailing off for a few days in the ship's boats to explore the region.

Darwin was convinced that Fitzroy's experiment with the Feugians had no chance of success, but he kept his thoughts to himself. He viewed the

Fuegians as miserable savages. Jimmy Button had told him that the men killed and ate women during the winter when food was scarce, and the ship's surgeon observed a child drop a basket of sea gull eggs, whereupon the enraged father picked him up and swung him repeatedly against the rocks, then abandoned him to die.

Ten days after leaving, Fitzroy, Darwin, and the sailors returned to the Fuegian village. Matthews, the young minister they had left behind, was almost beside himself with relief at seeing them. No sooner had Fitzroy and his men sailed away than the Fuegians attacked him and stole all his clothes and supplies. He feared for his life every moment they were gone. The natives had also turned on the three Fuegians who were dressed in European clothing.

Fitzroy was devastated and could not understand why the Fuegians had reacted so to his efforts to help them. Matthews was taken back aboard the Beagle when it sailed, but the three educated Fuegians were left behind in the hope that they still would have some success at spreading the gospel of Christ. A year later when Fitzroy and the Beagle returned to check on them they found that Jemmy Button, York Minister, and Fuegia Basket had discarded their European clothes and ways and were now indistinguishable from the other natives.

Darwin - Watercolor by George Richmond. 1840

Like so many college students today, young Charles Darwin could not decide what he wanted to do with his life. His father was a physician, and although the family was not wealthy, they were people of means and he grew up with privileges denied the majority of young people in England. At his father's urging he studied medicine, and when he discovered that he didn't have the stomach to be a doctor, he studied for the clergy. Like young people today, Darwin was acculturated by his times. Deeply religious, and with a Christian view of the world, there was nothing about him that indicated the path his life experience would take.

When Darwin graduated at the age of twenty-two, he seemed to be a very ordinary person: in twenty-two years he had done nothing that would distinguish him from other people. His grades were well below average throughout his college career, but he was an outgoing young man who made friends easily and despite his mediocre academic achievements, he impressed several distinguished professors who became his mentors. There was one thing about Darwin that caused him to catch the interest of some of his professors: he loved to collect things, mostly natural history things. Beetles, insects in general, lizards: he had them everywhere in his room. He also loved to hunt and to shoot. It may have been these interests that caused one of his professors, a man named Henslow, to call him aside one day and tell him that a ship was leaving to chart the waters off South America, that a naturalist was needed to go along, and that he had recommended Darwin to the naval authorities.

It was common in those days to include a naturalist aboard naval vessels to collect specimens and send them back to England to be identified, catalogued, and classified. Much of the world was still unexplored, and every time a ship went out it returned with hundreds or thousands of species of plants and animals that were new to the biologists of Europe.

Before being accepted to the position, Darwin had to be interviewed by the Captain, Robert Fitzroy. Darwin was very excited about the prospect of a sea voyage of exploration, partly because he wanted to get away from his father. He learned on the way to be interviewed that others were being considered for the position and that all of them were much more experienced than he as a naturalist.

Robert Fitzroy was very different from Darwin. He had entered the naval service at the age of 14 and was from a wealthy, aristocratic family. At the age of 23 he became captain of the Beagle, a ten gun ship with a crew of 74 aboard. His life up until that time contrasted sharply with the irresolute life Darwin had led. Against all odds Darwin and Fitzroy hit it off extremely well, so much so that even before the interview had ended the captain decided that he wanted Darwin and was trying to talk him into taking the position.

It was Fitzroy's decision to take a naturalist along on the voyage, and he had a personal reason: he believed that every word in the Bible was absolutely and literally true, and he wanted a naturalist along to find evidence for the historical validity of the Bible. In particular he wanted to find evidence of the great flood and confirmation for the book of Genesis story of creation. This sort of inquiry was not uncommon in early 19th century Europe. Most scientists were more properly natural theologians who thought the way to understand and honor God was to study His creation, both the physical world and the living. Remember that Darwin too had these same thoughts. He was to be a clergyman, a country parson.

"In order to pass the B.A. examination, it was, also, necessary to get Up Paley's "Evidences of Christianaity" and his "Moral Philosophy". The logic of this book and as I may add of his "Natural Theology" gave me as much delight as did Euclid. The careful study of these works…was the only part of the Academic Course which, as I then felt and as I still believe, was of the least use to me in the education of my mind. I did not at that time trouble myself about Paley's premises; and taking these on trust I was charmed and convinced of the long line of argumentation."
- Charles Darwin. Autobiography

That Darwin was an advocate of natural theology prior to his voyage is an important point, so I will digress for a few paragraphs to discuss the concept. William Paley (1743-1805), to whom Darwin referred in his autobiography, was an Anglican minister who wrote several books on philosophy and Christianity which were extremely influential in scientific circles and with the public. His most influential book, *Natural Theology: or, Evidences of the Existence and Attributes of the Deity, Collected from the Appearances of Nature*, was first published in 1802. In it he espoused the belief that the nature of God could be understood by studying His creation, the natural world. To illustrate his belief he came up with a metaphor that is often cited today to support the concept of intelligent design:

…when we come to inspect the watch, which we perceive…that its several parts are framed and put together for a purpose, e.g. that they are so formed and adjusted as to produce motion, and that motion so regulated as to point out the hour of the day; that if the different parts had been differently shaped from what they are, or placed after any other manner or in any other order than that in which they are placed, either no motion at all would have been carried on in the machine, or none which would have answered the use that is now served by it…the inference we think is inevitable, that the watch must have had a maker – that there must have existed, at some time and at some place or other,

an artificer or artificers who formed it for the purpose which we find it actually to answer, who comprehended its construction and designed its use. The marks of design are too strong to be got over. Design must have had a designer. That designer must have been a person.

Paley believed that if God had taken such care in creating other organisms, how much more He must care for human kind: "The hinges in the wings of an earwig, and the joints of its antennae, are as highly wrought, as if the Creator had nothing else to finish. We see no signs of diminution of care by multiplicity of objects, or of distraction of thought by variety. We have no reason to fear, therefore, our being forgotten, or overlooked, or neglected."

Paley's arguments were not new. Earlier writers like John Ray and others had proposed similar concepts, but Paley was widely read in the colleges and universities, and many of the scientists of the day embraced his philosophy. Darwin was one of those, as the earlier quote from his autobiography illustrates.

Now, back to the voyage. Darwin was beside himself with excitement at being offered the job aboard the Beagle, and he rushed home to tell his father. His father quickly dashed any hopes he had of leaving home. He was to stay at home and be a country parson. Darwin was crushed. His hope of adventure and independence from his father would not be possible. Very reluctantly he wrote to Fitzroy declining the offer.

The next morning while hunting with his wealthy uncle, Josiah Wedgewood (of the Wedgewood china family), Darwin told him the story, and his uncle interceded with his father and convinced him to let Darwin go on the voyage. And so a long series of events was set into motion that would result in a scientific theory that would one day forever change the way we think about ourselves and the world. It came close to never happening at all. Had Wedgewood not interceded, or had Darwin's father held firm in his opposition to the trip, Charles Darwin would almost certainly have spent the rest of his life as a country parson. Was it chance, fate, or divine intervention? Whatever the cause, we can easily imagine the excitement of the young man at the prospect of his first sea voyage, and more importantly, the opportunity to be on his own for the first time in his life.

Most people of Darwin's day firmly believed that the earth was unchangeable and relatively young: that all of the plants and animals on it were unchanged since the beginning of time. An English archbishop named Ussher had determined from studies of Biblical texts that the earth was created at 9am, on Sunday the 23 of October, 4004 BC. We can ridicule Ussher's calculations today, but they were widely accepted in Darwin's day, even among some scientists. For a theory of evolution to ever gain acceptance it was necessary for the extreme age of the earth to be accepted as well as its extraordinarily dynamic nature. The belief that the earth was only 6000 years old and had not changed since its creation negated any possibility that species of plants and animals could have evolved to their present form. The way Europeans looked at the earth began to change in the sixteenth and seventeenth centuries with the discovery of whole new worlds. Suddenly the old order of things was being challenged. Suddenly the horizons were so much greater. Suddenly there were new things to stimulate the mind – new animals, new plants, strange people, and unheard of things coming back to Europe from places no European had ever visited before.

It is interesting to note that the fixed hierarchical order of Biology began to pass with the disappearance of feudal society and the violence of the French revolution. In fact, the first serious evolutionists were French – Compte de Buffon (1707-1788), Georges Cuvier (1769-1832), and of course, Jean-Baptiste Lamarck (1744-1829). Lamarck proposed a widely accepted mechanism of evolutionary change in species that relied on the inheritance of characteristics acquired in the life of the organism. His well-known example of a Giraffe stretching its neck while trying to reach leaves higher on the trees during dry seasons, then passing the longer neck on to its offspring, was later shown to be flawed since a stretched neck can no more be inherited than a wooden leg or an artificial hip.

Prior to Darwin's sailing events were also occurring in England that were changing the temper of the times. Darwin's own grandfather, Erasmus, had proposed a theory of evolution which had gained some acceptance, and a new geological theory about the age and mechanism of geological change was radically altering current concepts. These would greatly influence Darwin's view of the natural world. Despite all of this, however, at the time he left England Darwin did not question the concept that species were fixed, meaning that they had never changed in the past and were not changing in the present. According to this view all of the species present at the creation were still present and no new species had arisen.

When on Board the Beagle I believed in the permanence of species, but as far as I can remember vague doubts flitted across my mind.
- Charles Darwin, 1877

Darwin's belief in the permanence of species was in agreement with what the majority of scientists of the day believed. The concept of species was formally codified by the 18th century biologist Linneaus, whose system of classifying living organisms into groupings ranging from large to small (kingdom, phylum, class,, order, etc), and assigning to each a genus and species name, solidified the concept of species permanence and invariability.

It is interesting to note that despite what appeared to be a general recognition that species were fixed, all of the separate pieces of Darwin's later theory of evolution were being discussed within the scientific community. Scientific theories are not born in a vacuum; they are a product of the culture of the times. The culture of Darwin's time was ready for a theory of evolution. What was needed was not more information so much as a person capable of synthesizing what was known into a single, grand, all encompassing theory. Most of the scientists of Darwin's day and those that preceded him who were capable of making the transition from natural theology to a unified theory of evolution were either too timid to do so or were simply unable to throw off the accepted beliefs of a lifetime and strike out into new territory.

What people do in times of change is to compromise, to change the old ideas just enough to make them fit the new information or situation. That's what Fitzroy wanted to do. He wanted to look for compromises that would allow him to accept the literal interpretation of the Bible regarding the creation of life. It's what Darwin initially wanted too, which is one of the reasons he and Fitzroy hit it off so well.

Somewhere in the course of the voyage Darwin reached the point where he was no longer willing to compromise. That was a tremendous obstacle for the young man to overcome. In England at that time the acceptance of the Biblical interpretation of the origin of life was more or less absolute except for a small minority of people. Darwin was a neophyte naturalist with minimal training in biology; certainly not someone who would be expected to challenge strong societal beliefs.

One of those scientists who almost made the transition from natural theology to a unified theory of evolution was a geologist, Charles Lyell. He came very close to proposing a theory of evolution not much different than Darwin's, but he drew back. For whatever reason, he just could not make the step. Lyell was a disciple of James Hutton (1726-1797), a Scottish geologist often recognized as the father of modern geology and as the man who established deep time. Hutton dispelled a view of the earth that had been a bedrock of Christian theology for nearly 1800 years – that the Earth was only a few thousand years old, and that changes since the creation were due to great catastrophes like Noah's flood. To Hutton, the Earth was dynamic and "alive", and constantly renewing itself as the forces of erosion – wind, rain, glacial movements – wore away the old surfaces. His studies of the rock layers near his farm led him to believe that the world is a place where the destructive forces of nature are counterbalanced by powerful forces which uplift new lands even as others are being worn away. More importantly, he believed that the same forces that create and destroy the surface of the earth in the past are still operating in the present. His theory of slow, gradual change over long periods of time is referred to as uniformitarianism. Despite the fact that Hutton was a devout Christian who believed that the Earth was created by God specifically for human habitation, he dismissed the Biblical story of a universal flood and other catastrophes as explanations for changes in the Earth's surface.

Hutton believed that his conclusions were an extension of his faith, not a break with it. To his mind, God created a world that was self-sustaining and self-perpetuating, literally without beginning or end. In this he went so far as to believe that the age of the earth was infinite. It was only centuries later when reliable dating techniques were developed that an age of about 4.5 billion years was arrived at by scientists.

It is hard to overestimate the importance of Hutton's work to Lyell, Darwin, and the scientists of all later centuries, including the present one. He developed an image of a vital, active Earth of infinite age, an Earth with powerful creative and destructive forces that are constantly reshaping the surface of the planet in an endless cycle of change, and he paved the way for the acceptance of one of the most important scientific breakthroughs of the twentieth century, the theory of plate tectonics. Hidden in one of Hutton's tomes is the following remarkable statement:

"..if an organized body is not in the situation and circumstances best adapted to its sustenance and propagation, then, in conceiving an indefinite variety among the individuals of that species, we must be assured that, on the one hand, those which depart most from the best adapted constitution, will be the most liable to perish, while, on the other hand, those organized bodies, which most approach to the best constitution for the present circumstances, will be best adapted to continue, in preserving themselves and multiplying the individuals of their race."
- The Theory of the Earth, vol.2.

Hutton's prose was borderline unreadable even in his day, but if this incredibly long, convoluted sentence is carefully examined, it can be found to contain the kernel of Darwin's theory of natural selection: those individuals within a species which are best adapted to the prevailing environmental conditions will survive and multiply at the expense of those less well adapted; and yet Hutton was convinced that species were fixed since the creation. A modern concept of evolution was developing in the womb of the European scientific community, but no one was willing to give birth to it.

It remained for a Scottish disciple of Hutton's, Charles Lyell (1797-1875), to convince the scientific community that present day geological processes, rather than great catastrophes like the Biblical flood, can explain the geological upheavals in the Earth's history. When Darwin embarked on the Beagle he took with him the first volume of what was to become a three volume work by Lyell: *Principles of Geology*. In it, Lyell made the argument for uniformitarianism. The second volume came to Darwin while he was in South America and there is no doubt that he owes a great debt to Lyell for the development of his own theory.

The frontpiece of Lyell's first volume depicted the Temple of Serapis in Naples, Italy. Near the top of several of the stone pillars are dark rings made of holes drilled by mollusks. These rings mark changes in sea level. The temple was built above sea level, was later submerged under water, and finally was lifted once again above sea level. Since all of this happened in the short span of several thousand years, Lyell argued that similar geological processes could account for greater changes in the Earth's surface over much longer periods of time.

Fitzroy and Darwin, fast friends, and with great respect for each other, set sail and moved down the Atlantic to the coast of South America and Darwin's first great adventure – the tropical rain forests of Brazil. Darwin had never seen or even dreamed of forests so magnificent. Everywhere he looked he found new species of plants and animals by the thousands. His passion for collecting now had free rein, and he disappeared into the forest every chance he got. These were not short one day jaunts; indeed, they were often trips of 500 miles or more through unexplored tropical forests. He traveled alone or with a small group of locals that he hired to accompany him. His life in England had been very soft, and he wasn't in shape for the hardships he encountered, but he pushed himself relentlessly and collected specimens by the barrels full. Hundreds of species of birds, thousands of species of insects; the young man was almost beside himself with excitement at what he was seeing and collecting. Nothing in his whole life had consumed him like this. Gradually he became inured to the hardships and his body became strong and capable of the most intense exertions.

The ship moved on to the pampas and he found there rock beds containing the fossils of huge animals that no longer could be found on the Earth, animals like giant armadillos and sloths, and he got into his first spat with Fitzroy by asking him how it was possible that such gigantic creatures as these could have been taken aboard the Ark. What most intrigued Darwin about the fossils, however, was that they so resembled the much smaller counterparts of these animals still alive in the South American forests. Even more interesting were the fossil remains of horses. When the Spanish first arrived in the Americas there were no horses to be found anywhere. Here was evidence that horses had existed in South America in ancient times and had become extinct there.

Children of God: Children of Earth

Fitzroy came to him one day and wanted to know where all this was taking him - where are you going with this Darwin? And how far have you succeeded in relating these things to the fundamental truths of the Bible? Darwin was noncommittal, but he had completed his reading of Lyell's first volume on uniformatarianism and his interpretation of natural events was beginning to change.

H.M.S. Beagle at Tierra del Fuego. Watercolor by Ray Massey

The Beagle left Tierra del Fuego and sailed up the west coast of South America and Darwin got his first glimpse of the Andes. The ship had no sooner landed than he hired several natives, acquired mules, and headed into the mountains to investigate. He was gone for six weeks. At 12000 feet elevation he came across fossil sea shells in great abundance. He immediately recognized that these rocks had formed under the ocean, now 700 miles away, and then been uplifted 12000 feet in elevation. The mere thought of the magnitude of the forces great enough to raise those majestic mountains to such heights greatly excited his imagination, but he realized that not everyone would share his insight or excitement.

We see here a quantum change in the young man's thinking since he left England: the Earth is not static, it can and does change. Mountains can rise, not catastrophically, but gradually over millions of years. Inch by inch and foot by foot what is seafloor can rise to the heights of great mountains. Fitzroy rejected this entirely. To him these were the mighty works of God, unchanged since the creation.

About a month later as Darwin visited the town of Valdivia on the coast of Chile an enormous earthquake rocked the town. Wide fissures opened up and closed again. The earth Darwin was standing on was writhing and jumping. It was terrifying. The quakes caused a series of tsunamis which completed the destruction initiated by the tremors. The epicenter of the quake was further north, near the city of Talcahuano, and that city was completely destroyed. .

Darwin was appalled by the destruction and loss of life, but he also noted with excitement that the level of the land had risen several feet as a result of the quakes. He took Fitzroy aside and said, look, the land is higher than it was before the quake. If it can rise three feet in a matter of minutes, why can't it rise 12000 feet? And isn't this the very process that has produced those mountains? But, if it is going to rise only three feet at a time it will not produce mountains like the Andes in a few thousand years: it will take much longer than that. But Fitzroy, like most of the people of the town, believed the quake had been caused by God as a punishment for their sins. Their belief, like Darwin's, was a matter of personal choice, and no amount of argument would change either's mind. Each felt completely satisfied, intellectually and emotionally, by what they believed to be true.

Here, both in space and time, we seem to be brought somewhat near to that great fact – that mystery of mysteries – the first appearance of new beings on earth.
-Charles Darwin

On September 15, 1835, The Beagle dropped anchor off Chatham Island (now San Cristobal), the most eastern and oldest of the Galapagos Islands. These islands were, and continue to be, born of fire. All are of relatively recent volcanic origin, the oldest being about 4 million years old; the youngest is still being formed. The islands lie on the equator 600 miles west of Ecuador in the Pacific Ocean. There are 13 major islands, 6 minor islands, and 42 islets.

The Galapagos Islands are the result of hot spots in the earth's interior that burn through the crust and pour hot basaltic lava into the sea. Repeated eruptions eventually pile up and break the surface, forming an island. Since the island is sitting atop a tectonic plate that is moving westward (about 3 inches per year), it eventually moves beyond the hot spot and another island begins to form in its place. In this way new islands form in a chain as volcanic activity quiets on islands to the west. The Hawaiian Islands formed, and continue to be formed, over a similar hotspot in the same manner. As the newest islands are added to (there have been nearly 50 eruptions in the Galapagos since Darwin visited there), the older geologically quiet islands wear away from erosion by wind, water, and wave action.

The islands are a harsh and dynamic environment. When Darwin stepped ashore on the black, volcanic sand, the heat was so intense that it burned his feet through the thick soles of his boots. The islands got their Spanish name from the huge tortoises that live on them, some of which weigh as much as 500 pounds and stand chest high to a man when their neck is outstretched. Darwin caught three young tortoises and took them back to England. One of them is still alive as I write.

The islands have one of the strangest and most extraordinary floras and faunas of any place on earth. Along the shore basking on the black basaltic rock are swarms of dark iguanas which feed on seaweed and never venture more than 30 or 40 feet from the water. A second species, the desert iguana, lives inland and scrambles down into the hot cauldrons of volcanoes to lay its eggs. Flightless cormorants, geckos, snakes, penguins, an albatross, fur seals, two species of rodents, a hawk, mockingbird, 12 species of finches, a couple dozen other species of birds, and several hundred species of plants are endemic to the island and are found no where else in the world.

One wonders how all these organisms found there way to a group of small islands surrounded in all directions for hundreds of miles by open Ocean. Birds can fly, of course (or in the case of penguins, swim), but why would they leave their native habitats and move so far out to sea? Perhaps the earliest arrivals were migrating or were in flocks blown out to sea in a storm. Plant seeds could have arrived with the birds, either in their gut or carried in their feathers or in mud on their feet. The reptiles and mammals must have arrived on floating debris. Large rafts of floating trees are sometimes observed moving out to sea from rivers along the coast of South America. These could carry eggs or animals which would eventually wash up on the Galapagos Islands. By whatever means, plants and animals from South America did arrive to colonize the Galapagos, but they found there a much different, harsher environment than they were adapted to.

I have not as yet noticed by far the most remarkable feature in the natural history of this archipelago: it is, that the different islands to a considerable extent are inhabited by a different set of beings. My attention was first called to this fact by the Vice-Governor, Mr. Lawson, declaring that the tortoises differed from the different islands, and that he could with certainty tell from which island any one was brought. I did not for some time pay sufficient attention to this statement, and I had already mingled together the collections from two of the islands. I never

dreamed that islands, about fifty or sixty miles apart, and most of them in sight of each other, formed of precisely the same rocks, placed under a quite similar climate, rising to a nearly equal height, would have been differently tenanted; but we shall soon see that this is the case.
Darwin – "Voyage of the Beagle"

Darwin visited four of the islands and collected large numbers of plants and animals, but as he himself noted, he failed to record which island the specimens were taken from. He was later to regret this mistake. Each island had unique species, even though they were close together and relatively the same in terms of climate and geology. As the Vice Governor noted, the tortoises were different enough in appearance that he could identify which island they came from by sight. What he didn't note, is that at least some of these differences are adaptive to different conditions on the islands. On drier islands where ground vegetation is scarce, the front of the tortoises shell is arched up high to allow the animal to reach up for leaves on the widely scattered bushes. On wetter islands where ground vegetation is more plentiful, the animals eat with their heads down and the shell is not arched in front.

…one might really fancy that from an original paucity of birds in this archipelago, one species had been taken and modified for different ends.
- Charles Darwin – From his diary

The thirteen species of finches he collected were of particular interest to Darwin after he reached England and began to catalogue his specimens. It was then that he regretted not labeling the specimens as to which island they came from. These small, nondescript birds have become famous the world over as "Darwin's Finches." The most conspicuous difference between the species is in body size and the size and shape of their bills. They also differ in their foraging behavior: some feed on seeds on the ground and have relatively large bills capable of crushing the seeds they find there; while others have smaller, blunter bills used for feeding on vegetation above ground. One famous species uses a spine to dig insect larvae out of wood after the manner of woodpeckers.

Darwin postulated that all thirteen species had originated from a single ancestral species which had migrated from South America. This one species then evolved into the thirteen species he found living there. Fitzroy's comment on this when Darwin spoke to him of it was that this appears to be one of those admirable provisions of infinite wisdom by which each created thing is adapted to the place for which it was intended.

Darwin's hypothesis regarding the origin of the Galapagos finches has until recently never had any evidence to back it up. Evolutionary Ecologist Ken Petren, working with Peter and Rosemary Grant of Princeton University, a husband and wife team that have worked for many years studying the finches in the Galapagos, used DNA analysis to compare the Galapagos finches with each other and with finches from the South American mainland. Based on the results of their studies they have concluded that all thirteen species had a common ancestor, and that ancestor was a warbler finch from the South American mainland. It would appear that Darwin was correct in his assumption as to the evolutionary origin of the finches. The young man had come a long way in accepting that species change with time and that new species can be created out of existing ones.

As Darwin became more and more convinced that species change in response to their environment, and that this can lead to the formation of new species and the extinction of others, Fitzroy became more and more rigid in his Biblical views. One can just imagine the two young men in the captain's tiny cabin, a dim lantern swaying gently above them with the roll of the ship, Fitzroy with his Bible and Darwin with his specimens, each making impassioned arguments to support his case. These arguments were important in helping Darwin sort through his own thoughts and build a case for the kind of evolutionary change he was observing almost daily. Both men were in agreement on one point: everywhere they went they saw plant and animal species specifically adapted for that locality, whether it was the Galapagos Islands, the Brazilian rainforests, or Tierra del Fuego.

The difference between them was that Fitzroy accepted that the world was only a few thousand years old and that it had not changed since it was created, while Darwin was now convinced that the world was many millions of years old and never stopped changing. Darwin reasoned that if plant and animal species are always well adapted to local conditions, and if those conditions are constantly changing as Lyell had so convincingly proposed, then the species must be constantly changing to remain well adapted. A concept of species immutability in a changing environment is a contradiction that will not stand up to examination. Obviously if the environment changed and the species were fixed, they would no longer be adapted to the local conditions. Darwin was more or less forced to accept by what he was seeing in his travels that species change and new species arise in the course of time while others become extinct.

The rest of the trip after the Galapagos was simply an effort to get home to England. They had done most of the survey work they were required to do except for making some chronological sightings. The Beagle made stops at Tahiti, New Zealand, Australia, Tasmania, and other places. Darwin had very little opportunity to study the extraordinary animals of Australia, but his comment after seeing a platypus was that "…two distinct Creators must have been (at) work." The Beagle sailed around the southern tip of Africa, crossed the Atlantic to the coast of South America to pick up the Gulf Stream, and sailed for England. They reached there on October 2, 1836.

CHAPTER 4

What Hath Darwin Wrought?

There is a grandeur in this view of life, with its several powers, having been originally breathed by the creator into a few forms or into one; and that....... from so simple a beginning endless forms most beautiful and most wonderful have been, and are being evolved.

- Charles Darwin

In June, 1860, two years after the publication of Darwin's book, a meeting was held at Oxford University which brought together supporters of Darwin and religious leaders to debate the theory of the origin of species. It is somewhat difficult today to imagine the alarm with which the clergy of 1860 viewed the growing acceptance of Darwin's theory, but they properly understood that their whole homocentric picture of the world was being seriously threatened, and they came to the meeting determined to crush Darwin and his theory. They were led by the Bishop of Oxford, Samuel Wilberforce, whose influence in the society of that day went well beyond the church.

Darwin declined to attend the meeting, but several of his staunch supporters were there, including Thomas Huxley, sometimes referred to as "Darwin's Bulldog" for his aggressive defense of Darwin and his theory. The first two days of the meeting dragged on without controversy, but on the third day Bishop Wilberforce was scheduled to speak and excitement was high in anticipation of his appearance. So many students and other people showed up that the meeting had to be moved to a larger room.

The Bishop was an imposing figure in his priestly robes as he rose to speak. His overly affected manner of speaking had earned him the nickname "Soapy Sam", and he was in fine form this day as he passionately attacked and ridiculed Darwin and his theory as being contrary to all common sense and the divine word of God as revealed in the Bible. At one point he turned to Huxley and in an imperious voice demanded to know if it was through his grandmother or his grandfather that he was descended from apes.

Thomas H. Huxley. 1874.

Huxley was both amused and appalled at the lack of any real substance in the Bishop's presentation, and in a loud aside said, "The Lord hath delivered him into my hands." He rose and announced that he would rather be descended from an ape than from a cultivated man who prostituted the gifts of culture and eloquence to the service of prejudice and falsehood.

In other words, Bishop Wilberforce didn't know what he was talking about. the Bishop's insolent question was probably not planned, but it was a mistake to pose it. Huxley was known as a staunch defender of evolutionary theory and of Darwin. Sometime during the mid eighteen seventies he remarked to a student, Henry Fairfield Osborn, later the director of the American Museum of Natural History, "You know I have to take care of him-in fact, I have always been Darwin's bull dog."

Children of God: Children of Earth

Cartoon from Vanity Fair of Wilbeforce during the debate

There was a moment of stunned disbelief that Huxley would so demean a man of the Bishop's standing; then uproar ensued, with clergy and Darwin supporters all trying to be heard above the noise. Amid the uproar a slight grey-haired man got to his feet, and in a voice shaking with rage, waved a Bible above his head. The truth was to be found here, he cried. Here and nowhere else. Years ago he had warned Darwin to give up his dangerous thoughts. He tried to continue, his face purple with rage – If he had only known what the result would be he would never have allowed – but his voice was lost in the noise of the crowd. The man was a Vice Admiral in the Royal Navy, and his contributions to the science of navigation are still recognized today, but on that day he was chocking on bitter bile, because it was he who had taken the young Darwin aboard his ship the Beagle and thus made it possible for him to challenge the Vice Admiral's most cherished beliefs. Robert Fitzroy fled the room, but the anger and guilt which consumed him on that day never subsided.

Fitzroy must have realized that his own contribution to Darwin's theory went beyond merely providing the travel experiences from which Darwin had conceived his theory. He had actively aided Darwin by constantly arguing with him, thus helping him to shape and reshape the amorphous ideas swirling in the young man's head. There was no getting around it; Fitzroy was as important a contributor to Darwin's theory as Lyell or Hutton. The thought apparently became too much to live with. On a Sunday morning in 1865, he committed suicide, still ruing the day he had chosen Darwin to accompany him on that now famous voyage of exploration aboard the Beagle. One has to wonder if in those few moments before he took his life his thoughts ranged back to those endless, sun-filled days thirty years earlier, when two young friends sailed the world in the greatest adventure of their lives.

The specimens and writings which Darwin sent home during the course of the five year voyage of the Beagle made him a celebrity in scientific circles long before he returned home. After a short trip to visit his family, he took up residence in London and immediately began to sort and catalogue his specimens, shipping many of them off to other scientists to work on. He completed this enormous task by 1846 except for a barnacle the size of a pin head. Most people at that point would have thrown the barnacle out the window or set it aside for good, but Darwin spent the next eight years studying barnacles and published 8 large volumes on them.

Darwin never again left England, never had a regular job, and was never again completely well. He had been seasick during the entire five year voyage of the Beagle, which may have been the cause of his ill health, but medical experts who have studied the symptoms of his disease think he may have suffered from Chagas's Disease, a parasitic infection transmitted by an insect common in certain areas of South America. From 1837 on he suffered episodes of stomach pains, vomiting, severe boils, palpitations, trembling, and other debilitating symptoms, none of which responded to the medical treatments of the day. In July, 1837, he began a secret notebook on transmutation of species which led to the development of his theory of evolution by 1838. Knowing full well how Victorian England would react to his theory, he confided only in close friends like Huxley and Henslow.

On January 29, 1839, Darwin married his cousin Emma Wedgewood in an Anglican ceremony. He had confided his views on evolution to her prior to their marriage and she apparently felt uncomfortable with them, writing to him to read from the Gospel of St. John "*If a man abide not in me, they are burned.*" Three years later they moved to Down House, a former parsonage in Down, Kent where he lived and worked for the remainder of his life. Being wealthy from his and Emma's inheritance as well as from the profits of his books, he never held a regular job but worked tirelessly on his research and writings, eventually publishing nineteen books.

Darwin was not a radical who wished to change the world. He was, in fact, a rather shy, retiring person who wished fervently to avoid controversy. He only rarely attended public meetings and never defended his theory in public, leaving that to his friends and supporters. Writing to a friend he once commented that admitting that species can change was like confessing to a murder. Never far from his mind was what he had observed during his trip. Living organisms change with time in response to environmental change; new species arise while some species become extinct. That species change was not a new idea. The question Darwin puzzled over was the mechanism of change. From his observations it was obvious that living things don't change at random; they change in ways that enable them to survive environmental challenges. How is this possible? By 1844 he had written a 240 page version of his ideas on natural selection, but fearing a public outcry that would ruin his reputation and social standing he held off publishing it.

In June, 1858, Darwin was shocked to receive in the mail a paper from a young English naturalist working in Indonesia. The man was Alfred Russel Wallace (1823-1913), and his paper, *On the Tendencies of Varieties to Depart Indefinitely from the Original Type*, proposed a theory of evolution nearly identical to Darwin's, which Wallace claimed was developed one night during a malarial fit. Despite having worked on his theory for nearly 20 years, Darwin was inclined to concede precedence to Wallace. Huxley and other friends who were aware that Darwin had formulated his theory while Wallace was still a teenager, submitted some of his writings on the subject along with Wallace's paper to the Linneaan Society where they were read concurrently but attracted little public attention.

Darwin had been working for years on a large tome explaining and defending his theory. He now put this aside and began work on a shortened version which was published in November, 1859, under the title *On the Origin of Species by Means of Natural Selection*. The first edition of 1,250 copies sold out on the day the book was released. In the book Darwin deliberately avoided alluding to his belief that evolution applied also to the human species except for a statement that "light will be thrown on the origin of man and his history," but he later published two books, *The Descent of Man* (1871), and *The Expression of the Emotions in Man and Animals* (1872), which dealt with the subject.

Toward the end of his life Darwin achieved the social acceptance and regard he so desired. He died at Down House on April 19, 1882, at 73 years of age. His last words, spoken to Emma, were, "I am not in the least afraid to die." Darwin wanted to be buried in a plain, unadorned coffin in a churchyard cemetery in Down where he had lived most of his adult life. His scientific colleagues felt that a greater honor was due him and they arranged a state funeral and burial in Westminster Abbey next to Sir Isaac Newton. It is ironic that the man who, in the words of Loren Eisely, "shook the foundations of Christianity," should be buried in one of the oldest and most revered shrines in the Christian world.

Natural Selection

So, what hath Darwin wrought? His concept of natural selection is one of the most powerful tools ever developed for understanding the natural world. It is the bedrock upon which all of modern biology rests. His theory initiated an explosion of progress in the biological sciences that transformed the study of living organisms from a disparate group of loosely knit and often antagonistic disciplines into an integrated field of study with a pace of discovery that continues to increase today with phenomenal speed. That much cannot be disputed by thinking people. But what has it meant in human terms? How has it changed the way we think about ourselves and the world around us?

Even the staunchest detractors of Darwinism today must agree that because of him very few people still think of the world, or our place in it, in static terms. We have accepted so completely that change is the essence of the world, including our personal world, that the static world-view of the middle ages is nearly incomprehensible to us today. Why then, in one area of thought – our own biological history as a species – do so many people balk at Darwin's theory? Even people who accept that we are a species with an evolutionary history that extends back to the lowliest creatures on earth often feel uncomfortable with the thought.

I believe it is all about control. The human species has striven throughout its evolutionary history to wrest control away from the external environment and away from the gods of our ancient ancestors. In terms of seizing control for ourselves the greatest change in human history was the development of agriculture and the consequent disappearance of a nomadic lifestyle. Food and water are the most important commodities for survival, and seizing control of the food supply from the gods of nature was the beginning of our supremacy as a species. Agriculture allowed people to settle in one place and begin to build cities, with permanent buildings to protect against the vagaries of the environment. It enabled them to produce a single crop in large amounts and to store the food against the winter. It allowed for cottage industries to develop to produce clothing and metal tools. The surplus of food meant that not all people had to farm, and classes of artisans, politicians, tradesmen, craftsmen, became organized into what we think of today as a civilized society. The most important aspect of the society was the measure of control it gave people over their food supply and the environment.

The greatest change in human history since the Neolithic revolution with its development of agriculture was the end of Feudalism and the application of power-driven machinery to manufacturing. Closely associated with these changes was an agricultural revolution in England that by 1830 had transformed the countryside from an open-field system of cultivation to compact farms and enclosed fields.

A landmark in the industrial revolution was the development of the steam engine. This simple device radically altered a way of life at least ten thousand years old by replacing animal power with mechanical power. An incredible explosion of new machines followed.

Industrialization led to the death of cottage industries and the creation of factories, which in turn was largely responsible for the rise of modern cities. The British agricultural Revolution made food production more efficient and less labor-intensive, forcing the surplus population into the cities to seek work in the newly developed factories. The factories were able to turn out goods quickly and cheaply, making it possible for people of all classes to enjoy an improved quality of life.

The Agricultural and Industrial Revolutions in England empowered people like nothing had since the development of agriculture and the disappearance of a hunting-gathering existence. Food, clothing, household goods, transportation, and fuel – all were readily available in quantities unimaginable before, and the population grew in leaps and bounds.

The Agricultural and Industrial Revolutions were preceded and accompanied by the Renaissance which brought about an artistic transformation and scientific revolution in Europe. During the 17th century Kepler, Galileo, Newton, and others developed mechanistic models of the solar system and universe which they felt clearly showed the exquisite beauty and organization of God the Creator. The natural theology of Hutton and Lyell extended this thinking to the geology of the Earth, as did the work of Linnaeus, Cuvier, Lamarck, and others to the realm of living organisms. It was all so perfect and comforting because it reflected and highlighted the perfection of God.

Then came Darwin, and the whole edifice of natural theology came crashing down. It was all so sudden. His book on the Origin of Species hit the public and the scientific community like a bombshell. They were not prepared for it, and their confusion turned to anger and ridicule. Lifelong cherished beliefs were being threatened; indeed, God Himself was being threatened. Darwin had undergone a revolution in his paradigm of the world, but it had taken a five year trip around the world and two and a half decades of study to bring this about. The public and majority of the scientific community had none of these experiences to help soften the blow.

Darwin grew up in an era of rapid cultural change. In addition to the Agricultural and Industrial Revolutions new worlds were being explored and the Earth was suddenly much larger and more complex than anyone had ever dreamed. It was in this atmosphere of change that Darwin was able to visualize change in the natural world.

It is rare in science for a significant new theory to come full blown from whole cloth. That is not how science works. More often than not, a new theory is born only after all the elements required to support it have been brought forth by a variety of scientists over a fairly long period of time. The new theory is usually a synthesis rather than a brilliant new thought conceived in a vacuum. Such was the case with Darwin's theory of natural selection. The concept that long term changes could be produced by the operation of slow, ongoing processes was not Darwin's but Lyell's as set forth in his three volume work on uniformaterianism. Thomas Malthus (1776-1834) proposed the concept that living organisms produce more offspring than can possibly survive. Evolution as a concept was well established in scientific circles by the time of Darwin's trip, and several attempts to understand the mechanism of change had been proposed, the most influential of which was by the Frenchman, Jean Lamarck (1744-1829). Darwin's own grandfather, Erasmus, had also proposed a mechanism of evolutionary change.

Why then did Darwin succeed in putting the pieces of the puzzle together when other, more distinguished and established scientists like T.H. Huxley did not. (Huxley is reported to have said "How stupid of me not to have thought of it." when he learned of Darwin's theory of natural selection). A number of factors worked in Darwin's favor. He was a young man, not yet encumbered by the baggage of decades of traditional thought. His education was limited to that of an undergraduate degree, enough to discipline his mind and educate him generally, but not enough to codify traditional thinking. During his five year voyage he was isolated from all contact with the scientific community and his mind could range freely in all directions. He had the constant company of Robert Fitzroy who opposed his every thought regarding species change and thereby helped by acting as a sounding board for his ideas. Every time Fitzroy disagreed with him Darwin was forced to reexamine his thoughts and come up with new arguments. Darwin's enthusiasm for collecting led him into travels in the rain forest and mountains of South America where the profusion of plants and animals and their obvious adaptation to differing environments impressed the young naturalist immensely.

Darwinism and atheism are often linked together in the minds of people. They should not be. The texts of the Bible cover the creation of the physical universe and of mankind in two short chapters. What was the source of the ancient writer's knowledge about such things? Was it revelation from God, or as seems more likely, was it a recitation of oral tradition passed down from one generation to another from tribal societies long since vanished? They were men, just as we are men, and they tried as best they could to make sense of the world around them. They believed in God, as we believe in God, and lacking science as a way of knowing, they trusted to their own intuition and the stories of their animistic ancestors. We would have done likewise had we lived in their day, but that doesn't make what they wrote of value to us in interpreting the natural world.

Darwin's loss of faith in Christianity was not a result of his scientific work but of the personal tragedies he suffered, particularly the loss of his children. Upon leaving college it was his goal to become a clergyman. He was a firm believer in Christianity at the time, and fully accepted William Paley's arguments for the existence of God from the design of nature. But for an act of fate which put him on the Beagle, he would have spent his life as a country parson. While at sea he and Fitzroy together wrote a pamphlet urging more government support for the missionaries in the Pacific. The pamphlet was published after their return to England.

Darwin and Emma had ten children, three of whom died in childhood. The death of one of them, ten year old Annie, so affected Darwin that twenty-five years after her death tears would come to his eyes at the thought of her. With Annie's death Darwin lost his faith in a beneficent God. This is not an uncommon occurrence today with parents who lose children. In Darwin's case other factors were also at work to weaken his faith. He had been appalled at the savagery of the Tierra del Fuegians, and was horrified at the brutal, inhumane treatment of slaves which he observed in South America. The cruelty he observed in nature also contributed to his loss of faith in a kind and loving God. In keeping with his views on evolution, he eventually came to see religion as a survival tactic of the human species. On Sunday mornings when Emma and the children went to church he would go for walks until they returned. Deep down, however, Darwin never completely lost his faith. When asked about it, he always maintained that he was an agnostic, not an atheist.

It may well be that for posterity his name will stand as a turning point in the intellectual development of our western civilization... If he was right men will have to date from 1859 the beginning of modern thought.
    - Will Durant on Charles Darwin

Darwin's theory threw society into a contentious debate that continues to this day in the United States. According to his theory we are really the products of an indifferent creative process whose goal was never to create the human species. Natural selection is indifferent to us, a mechanistic process of creation that has no purpose other than survival. To many of the people of his time and ours his theory robs us of meaning and takes away all control over our past and future: we are an evolutionary anomaly – a two-legged creature of brief evolutionary history with a very questionable future.

We live fleeting lives in a very dangerous world, and it is comforting to believe that we are here for a purpose. If we are just one of millions of species competing for survival, then our special relationship to each other and the Earth is forever dissolved and our vulnerability to older, more primitive forms of life like bacteria, viruses, and insects – our nemeses as a species – becomes unnerving. They are remorseless, indifferent, cold – alien races, and in the end we know they will prevail. They have the ability to reduce humanity to insignificance. And so, we rationalize our superiority over them by denying our evolutionary ties and the debt we owe them for our very existence.

It also means that nothing stands between us and the violence of the physical world. Storms, earthquakes, floods, avalanches, mudslides, wildfires and other natural disasters can carry us off in the blink of an eye. Whereas we would prefer to think of ourselves in purely personal terms and believe that we are somehow special as individuals, the physical world is a constant reminder that our life is tenuous at best and that our survival in large measure lies in factors beyond our control

In our inability to accept the indifference of the world we elevate ourselves to the position of gods. The natural world is wild and uncontrollable, but it was created by a God who loves and looks after us, and therefore in the end we will rise above it all, leave the slime and muck of the earthly swamp behind, and reign for all eternity in a spiritual world that transcends all earthly harms.

Unity Church Prayer

I affirm peace, protection, healing, and divine order.
The love of God comforts and soothes you;
The light of God guides you and keeps you safe;
The life of God heals and renews you;
The power of God works through you to restore order
and rebuild your life.
You are sustained in body, mind, and spirit by the
ever-renewing presence of God.

It is perhaps understandable that we should object to a naturalistic view of ourselves, but are our objections realistic? Some theologians maintain that they are, that a theological view of human history presents a more rational and cohesive explanation than science. Without question giving purpose to the world is very comforting and explains much of what science cannot, but the theological viewpoint cannot be proved or falsified by science: scientifically, it is an intellectual dead end.

Darwin redefined our place in the world. His theory of Natural Selection not only identified a more accurate contextual relationship between our species and the rest of nature, but it also gave a breath-taking view of our evolutionary history and of the interplay of geological and biological forces that formed us. No longer were we created merely for the pleasure of an all powerful God. We now had an evolutionary history of our own – a history of struggle and failure, triumph and tragedy. We are survivors in a prolonged and titanic struggle for existence. We have out-competed our competitors for food, even though they were brilliantly designed for the competition. We have battled the deadly disease organisms – viruses, bacteria, fungi, parasites and others- and although millions have died, in the end we always triumphed.

Each of us stands at the end of a line of survivors which reaches back more than a million years. No, we are not the protected, pampered siblings of an all powerful God, and although we respect and worship Him, we know, largely because of Darwin, that we have made it on our own, and it should give us a sense of pride and security that is impossible within the Christian doctrine of special creation. Under that doctrine we are merely puppets whose every creation and movement is choreographed from on high. We have no self-determination, no right to an existence of our own.

Darwin's concept was immediately so compelling to some of his compatriots that the first edition sold out in a single day and a second edition was issued within a month. But there were equally powerful and passionate forces that saw their whole world view crumbling before them. They were people who could not admit to change, who could not stand to think that we had won the struggle alone. There was – and is – fear; fear that if we have struggled alone this far we must continue to do so, must draw upon our own resources to survive – alone.

The thought is terrifying to many people. It breeds a sense of isolation and purposelessness that they cannot endure to contemplate. Rather than seeing the strength of our position, they see only weakness and vulnerability. Their view of the world and of their place in it demands rigid structure and paternalism. Without these, they are lost souls in a dark and terrifying universe that is completely indifferent to their fate.

I have no argument with people like this except when they dogmatically demand that I and mine must be forced to think as they do. That battle still rages in the United States, just as it did in England after the publication of Darwin's book. Neither side is willing to concede and the debate is often fought out in the courts, textbook adoption agencies, school boards, and legislatures. It is a debate that has long since ceased to be productive. It has become mired in dogmatism, recrimination, and absolutism on both sides which saps our time and energy in a useless cycle of petty denunciation. It is not worthy of us.

CHAPTER 5

Evolution

(Evolution) is a general postulate to which all theories, all hypotheses, all systems must hence forward bow, and which they must satisfy in order to be thinkable and true. Evolution is a light which illuminates all facts, a trajectory which all lines of thought must follow.

- Pierre Teilhard de Chardin

Charles Darwin was not the first scientist to propose that living organisms change in response to changes in their environment and his theory was not entirely unique to him. Every postulate in his theory had been proposed in one way or another by other scientists. Darwin sifted through conflicting pieces of jumbled theories and pieced together a comprehensive synthesis that has withstood the strongest storms of criticism and controversy.

Once Darwin accepted that living organisms change with time, the question he wrestled with was the mechanism of change. How do living organisms change in a way that keeps them well adapted to a changing environment, and how can this lead to the formation of new species.

His answer to these questions is that nature selects from among the individuals in a population, favoring those which are the best adapted and eliminating those that are poorly adapted. His basic concept can by outlined in the following statements:

1. Every species strives to increase its numbers by producing far more offspring than can survive.

2. The resources of the environment which support a species are in limited supply.

3. Genetic variation within all sexually-reproducing populations produces an enormous amount of diversity within a species.

4. Some genetically determined traits within the population have greater survival value than others.

5. Individuals with favorable traits are at an advantage in the competition for limited resources and survival. These individuals live longer and leave more offspring, many of whom will carry the same favorable genes.

Let's consider each of these briefly. Every plant, animal or microbe is capable of producing enormous numbers of offspring. A mature walnut tree, for example, produces hundreds of walnuts a year. If all of these grew into trees and each of those trees produced hundreds of new trees per year, in a short time the world would be covered with walnut trees. This is clearly impossible because of space and resource limitations and because of competition with other species of plants.

This was a key element in Darwin's development of his theory. It came to him after reading an essay by Thomas Malthus (1766 – 1834) in which Malthus pointed out that the human population has the capacity to increase much more rapidly than its food supply. Populations under ideal conditions can increase exponentially, while an increase in the production of food lags far behind. Malthus's contention was that no matter how much the food supply is increased, a proportion of the human population will always live in poverty.

Exponential increases are awesome in their doubling power. There is a story about a young man who was not satisfied with his salary, so he went to his employer and offered to work the first day of each month for a penny if the employer would double the pay each day thereafter until the end of the month. The supposedly gullible employer readily agreed and paid the man a penny on day one, two pennies on day two, four on day three, eight on day four, sixteen on day five. By workday ten it had increased to $5.12 and on day 15 the daily pay had risen to $163.84. Doubling this each workday gave him $5,242.88 on the 20th day. This increased to $167,772.16 for a day's work by day 25 and to $5,368,709.12 on day thirty. Had he been fortunate enough to be working in a month with thirty one days, his pay for day thirty one would have been $10,737,418.24. His pay went from pennies to millions in just thirty days.

This is how a population of walnut trees, mosquitoes, mice, or any other living organism can grow under unrestrained circumstances. Two individuals (male and female) can become millions in less than thirty generations under the most ideal circumstance. For bacteria this may take only a few hours; for insects a few months; for birds or walnut trees, thirty years. Despite the awesome power of reproductive increase, most populations do not increase; most are in a dynamic equilibrium. How many offspring generally survive? On average, only two. If more than two survive, the population will grow; if less, it will decrease. The question then is, why are a very few able to survive while most do not live long enough to reproduce and pass on their genes into the next generation?

If all members of a species were identical genetically, the answer would be that survival is random and a matter of chance; however, except for genetic (identical) twins, no two members of a sexually reproducing population are genetically identical. Each is a one time, never to be repeated genetic event. No two humans in the entire history of the species were genetically identical unless they were twins from the same fertilized egg. You and I are one time genetic events that have never occurred before and will never occur again (My wife says this is a very good thing). You are unique, as are all living organisms except those which result from cloning. The explanation for this concept is explained in the chapter on sex.

This is a difficult concept to grasp, but it is absolutely essential to understanding and accepting natural selection. Darwin's genius was in providing a mechanism by which seemingly random genetic variation could be the raw material for adaptive change. The so-called insecticide treadmill will illustrate how the mechanism works. DDT was the first synthetic insecticide to be successfully employed to control harmful insects like mosquitoes and agricultural pests. It was first used in the late 1930's and to the delight of everyone concerned, it was wildly successful in killing a broad spectrum of insects. No one really knew how it killed, but they didn't care and it was produced by the millions of tons and sprayed everywhere, from farmers' fields to city streets and even down the clothing of displaced people to control body lice.

Farmers were beside themselves with joy. For the first time in history they had a weapon which could combat the insects which for millennia had been destroying as much as 40% of their crops in the field and in storage. They could spray a field one day and the field would be clean of insects by the next. DDT was a gift of miraculous proportions. Even highly respected entomologists thought that DDT was the ultimate weapon against insects and that we had finally won the battle against these ancient foes. They soon found out that they had been incredibly naive. Disturbing reports soon began to come in that DDT was losing its effectiveness. Farmers responded by increasing the dosage and shortening the period between sprays. This worked for a while, but again, the DDT was not as effective as it had been initially. The same pattern occurred when other synthetic insecticides like dieldrin were introduced. For a while they would be extremely successful, but as time passed their effectiveness dropped dramatically.

What was happening? The insects were adapting to an environmental challenge that threatened their existence. In a word, they were evolving. Insects have a short life cycle and a very high reproductive potential. Like humans and all other sexually reproducing organisms, every insect within the species is unique genetically. Among the insects in every field there existed a very small number which for *genetic reasons* were resistant to DDT. That resistance was present before their exposure to the chemical; the chemical *did not elicit it.*

The few insects which survived quickly reproduced and filled the field with their offspring. In doing so they passed on the DDT resistant genes, which meant that most of the insects in the field now had a higher level of resistance to the insecticide than earlier generations. When the farmer upped the dosage and shortened the time between sprays, most of the insects were again killed but a few survived that had an even higher level of resistance. These reproduced and filled the field with their own kind. Over time insects resistant to any concentration of the insecticide emerged, selected for by the insecticide.

The farmers were actually creating insects which could not be killed by a process of selection in the same way that an animal breeder produces better animals by breeding his best stock. Again, note that DDT did not create the genetic resistance, it merely selected for it by killing off all those individuals which did not have it in the first place. In a sense, this is artificial selection in that humans are the cause of it, but the exact same process operates in nature.

Why would insects have genes to protect them against chemicals like DDT, a man-made chemical which had never existed before? Plants have been waging a chemical war against insects and other plant eaters for millions of years. We think of plants as helpless. Not so; if they were, they would all be gone, eaten up by hungry animals. Plants can't run away from danger, they must stand in one spot, sometimes for centuries and "take it." They protect themselves with chemicals, and they are ingenious at devising them. Some plants produce chemicals that kill or sicken their predators. Some produce phony amino acids which are close enough to the real ones to be incorporated into the insects' proteins, but because the proteins will not function properly, the insect sickens and may die. Some plants produce a perfect copy of an insect hormone, juvenile hormone, which prevents the larval stage from becoming an adult. Since larvae lack reproductive organs, they cannot reproduce as long as they are feeding on the plant.

Insects have been successfully waging this war with plants for hundreds of millions of years. It is a back and forth struggle: plants come up with a new chemical which is highly effective for a while, but by the process of selection, the insects adapt to it and resume eating the plant. Later, through a chance mutation, the plant may again alter its chemical defense, which the insect must then once again adapt to or cease to exist. It is because of this ongoing war and the difficulty of adapting to a plant's defenses that the overwhelming majority of plant eating insects are forced to feed on only one, or at most a very few closely related, species of plants.

Nature selects the fittest (meaning best adapted) in each generation from the offspring produced by each reproducing pair of individuals in a population. No matter what species is examined – walnut trees, mosquitoes, robins, deer, lions, etc. – many more offspring than can be supported by the resources of the environment are produced in each generation. These are culled by natural selection, and in stable populations, on average only two survive to replace the parents. The two that survive do so because they are better able to feed themselves, escape predation, withstand the vagaries of the climate, survive the attacks of disease organisms, and, in general out-compete their peers for the resources of the environment. The process is far from random.

© *Tribune Media Services, Inc. All Rights Reserved. Reprinted with permission.*

Dog or Wolf?

The amount of variability within a species is a measure of its ability to adapt to changes in its environment: the greater the variability, the greater its ability to adapt. The sum total of all the genes present in all the members of a species or population is referred to as a gene pool. The amount of variability within a gene pool is usually very great. The wolf population is a good example.

Sometime around 12,000 years ago some of our ancient ancestors living in what is now Israel buried a body cradling a puppy. This is the earliest known fossil evidence of the domestication of dogs, but the practice is thought to have begun about 14,000 years ago. Dogs were domesticated from wolves, and from a genetic standpoint they are still wolves. The DNA makeup of dogs and wolves is nearly identical and all dogs are completely capable of interbreeding successfully with wolves. Naturally a wolf breeding with a Chihuahua would be impossible from a physical standpoint, but from a genetic standpoint there is no problem.

That a wolf, through selective breeding, could be changed into a dachshund, Border Collie, Bloodhound, Great Dane, Sharpei, or any of the other mind boggling array of shapes and sizes that dogs come in is a measure of the genetic variability within the wolves' gene pool. Not only do the 400 or so dog breeds look very different from wolves but their temperaments and behaviors differ in important ways also.

Dogs evolved in the company of humans and cannot exist without them. Our ancient ancestors domesticated wolves by selecting traits that helped them in their efforts to survive - protection, hunting, herding, and companionship. Wolves with these traits were selectively bred, leading to different breeds for each of these functions. The point, however, is that genes for all these traits – the Dachshunds short legs, great Danes' size, Bloodhounds' nose, Basset hounds long ears, sheep dogs herding behavior, etc. – were all present within the wolves gene pool before humans began to domesticate them. This is not natural selection since humans were performing the function, but the process of selecting favorable genes over unfavorable ones is the same whether performed by humans or nature. In both cases, the genes being selected were present in the gene pool before the process of selection began.

What is the result when a species comes up against an environmental challenge for which it has no genes? Extinction. A case in point is the fate of the American Chestnut Tree when it was exposed to a fungal infection introduced into North America from Asia. Originally one of the most common trees in eastern hardwood forests, it has been completely eradicated.

When genes are present to meet the threat, the population can rebound even though a major part is destroyed. An example of this is the Black Death which swept through Europe and Asia several times, killing as much as thirty percent of the population each time. To put this into perspective, imagine a disease sweeping through the United States and killing 100 million people in a relatively short period of time. Scientists are not sure whether the cause of the disease was Bubonic Plague or a viral infection, but many people who were exposed to the disease survived and were apparently immune to it.

Recently, Dr. Stephen J. O'Brien of the National Institutes of Health discovered a mutant gene (CCR5-delta32) which provides complete immunity to HIV when present in the homozygous condition (meaning both parents contributed a gene). Individuals with only one gene have a partial immunity. Approximately 10% of Europeans now have the mutant gene, but it is rare or absent in other parts of the world. Interestingly enough, according to Dr. O'Brien, the gene also provides immunity to the germ causing the Black Death, which accounts for the relatively high frequency present in European populations. Bubonic plague and the HIV virus both target white blood cells and enter them. The CCR5-delta32 gene prevents their entering. No matter how severe the HIV epidemic should become, or how helpless we were to deal with it, some humans would survive and repopulate the planet, but a high proportion of the rebuilt population would carry the once rare mutant gene. This is natural selection at work.

Dr. O'Brien pointed out that CCR5-delta32 is an unusual mutation in that it has no harmful effect on those who carry it. Most genes which mutate cause serious diseases like cystic fibrosis, sickle cell anemia or diabetes. In fact, most mutations are harmful because they are variants of highly successful genes. A mutation becomes useful only when the environment changes in some way and the previously successful gene no longer is adaptive. A classic example is a mutation which changes the molecular structure of the hemoglobin molecule and causes sickle cell anemia in which the normal dish-shaped red blood cells curl up into a sickle shape. As such, they cannot pass easily through the tiny blood vessels which feed organs and death results. Individuals homozygous for the mutant usually die from the disease. Those with only one copy of the gene can suffer serious harm which can lead to death and so the gene remains at very low levels in most of the populations of the world. However, in parts of Africa where malaria is rampant, the sickle cell gene occurs with a high frequency because it confers some immunity against malaria. Approximately 100,000 people die yearly from sickle cell anemia while 1,500,000 die from malaria. Obviously, selection is for the gene conferring the best chance for survival, even though that gene is considered to be a harmful one.

Microevolution, as this process is called, has been grudgingly accepted by all but the most dogmatic of Christian fundamentalists. A few diehards contend that it is not evolution because nothing new was introduced into the population: both the normal gene and mutant gene were present prior to the change; the only difference is in their relative frequencies. That is quite true, but genetically and in its response to the environment, the population is different, and that difference is critical to the survival of the population or species. If none of the people living in Europe and Asia had carried a gene which resisted the attack of the Black Death, it is quite possible that they would have gone the way of the American chestnut tree. This almost happened to Native Americans. Ninety percent of them were killed by common European diseases like small pox and measles because they had no resistance to these diseases which were new to them. Our species is not above natural selection.

Selfish Genes?

In a controversial and highly popular book published in 1976 entitled The Selfish Gene Richard Dawkins proposed that the gene, not individuals or populations, are the unit upon which natural selection acts. According to this concept, genes act "selfishly" to increase the number of copies of themselves and individual organisms exist only to propagate their genes. In Dawkin's words, organisms are only the "vehicles" or "survival machines" for their genes. The old saying "A chicken is just an egg's way of making more eggs" pretty well sums up the concept.

Although I do not much care for this view of living organisms, one aspect of it does appeal to me. Individual organisms are mortal, but genes are in a sense immortal in that they pass from one generation to another without any discontinuity. Because, as Dawkins points out, genes flow like a river through time while individual organisms are mere interlopers whose existence passes in a microsecond of geological time, focusing on the gene helps to appreciate the timeless nature of evolution.

Dawkins could just as well have focused on the gene pool of a species or population rather than on individual genes. Using Dawkins concept a species gene pool could be thought of as a dynamic, ever changing, super society of genes in which robot-like organisms (humans, trees, etc.) are manipulated to achieve the society's extension in space and time.

Life as we know it is tiered, and each layer depends for survival on all the others. DNA, cells, tissues, organs, organ systems, organisms, populations, communities – all operate together in a coordinated and integrated way. A change at any level can have a cascading effect that will alter the way natural selection acts on the individual organism. Some of these changes could have nothing to do with the genes themselves. Focusing only on the gene as the unit upon which natural selection acts is an over simplification in my opinion.

Natural selection does not act only on genes but upon an integrated genome/environment/experience complex (GEEC). Genes individually and collectively represent only a potential. Human genes for height, for example, can fail to be expressed when a person is chronically undernourished. Genes for arms can fail to be expressed if a pregnant woman takes the sleeping drug thalidomide, resulting in an armless child. Organisms which natural selection act upon are more than the sum of their genes: they are the result of the interplay of environment, experience, and the genome (the sum of their genes). There is also a dynamic interaction of genes within the genome that can affect the expression of individual genes. The term "jumping genes" refers to genes which can move within the genome, often from one chromosome to another. Moving a gene from one place to another can alter its expression, as Barbara McClintock has shown. There are also genes that regulate other genes, and a mutation in one of these can alter the expression of the genes under its control. Individual genes do not act in isolation; they act in concert with the other genes that make up the genome.

The relative importance of the three factors in GEEC change with the life cycle and differ with different species. During development of an egg into an organism the genome and environment are controlling, and the influence of experience on natural selection is zero. Experience can remain at or near zero for species with very short life cycles and/or limited behavioral repertoires. I doubt that experience plays much of a role in the life of a sponge or an oak tree, for example. (Having said this, I believe we may be grossly underestimating the behavioral repertoire of plants.) For long lived species with a capacity for learning, behavior as a result of experience can mean the difference between life and death. Learning what foods to eat (or not eat), which animals to ignore and which to avoid, how best to attract mates, are critical to survival. Behavioral flexibility allows innovative behaviors that are adaptive to sudden environmental change. In long lived species with an expanded ability to modify behavior, experience will grow in importance with age.

The role of environment in survival following development is critical. Organisms all have limits of tolerance with respect to environmental parameters. When these are exceeded, the organism will fail. The most stable biotic systems can be changeable, even chaotic at times, but the role of environment as a test of an organism's ability to survive increases with the instability of the environment. When speaking of the role of environmental influences we must keep in mind that this includes both living and nonliving elements. A new species introduced into an organism's environment can be devastating for that organism's chances of survival. The destruction of the

American chestnut tree is a case in point. All environments are susceptible to change; this affects the way natural selection acts on the individuals living in them.

Individual genes are **not** selfish. They are prevented from being so by the reshuffling that goes on during reproduction. The shuffling of genes during the production of eggs and sperm is like a game of poker. No matter how good one hand of cards is, when the hand is over the cards are thrown together, shuffled, and a new hand dealt. Having a royal flush in one hand means nothing when the next hand is dealt. The value of individual cards is not as important as the combination that makes up a hand. In some instances a trey will be more valuable than an ace because of its relationship to the other cards (four treys beat two aces).

The same is true of genes. Combinations of genes act together in a way that is more than the sum of their individual actions. Gene combinations have multiple potentials. They are capable of interacting with the environment in a way that allows gene function to be modified to fit circumstances. If two oak trees with identical genes are planted in different habitats, one in a forest and the other in an open lawn, they will develop different growth habits in response to their environments. The tree in the forest will grow tall and thin as it reaches for the canopy and light. The tree in the lawn will grow low and squat, with many large branches low to the ground.

No matter how successful the combination of genes are in a particular individual organism, that combination will be broken up and new ones formed in the offspring which may have greater or less value in survival terms. How many great people have great children? Great male athletes like Pele, Michael Jordan, or Lance Armstrong may have children with a wide variety of athletic ability, none of whom are the equal of their fathers. Great scientists, musicians, artists, politicians, rarely have children who achieve greatness, but in each generation great scientists, musicians, artists, and politicians surface (well, maybe not politicians), often from parents without their children's natural abilities.

When my daughter was in third grade she announced to me one evening that she was going to try out for the school basketball team. I was incredulous that children had to try out for a third grade team. Children not making the team in third grade probably will never play at any higher level (at least in Indiana.) This process of selecting the kids with the most talent gets magnified all the way to the pros: several elementary schools funnel into a middle school which selects the best from each of their teams for its own team. Since the high school is consolidated, players from a number of middle schools try out for the team but only a relatively few make it. Out of all the high schools in the country a paltry few are recruited by colleges, and out of tens of thousands of college players a few dozen make it to the pros.

A process of selection that began in third grade among tens of millions of children culminates in a few hundred athletes of professional quality. Two things need to be noted about this process: genetically acquired ability, training, and environment (home, school, and community) are critical to the athlete's success; and, the survivors in this process were selected for out of a huge population. Coaches are fond of saying in one way or another that "I can't get out what God didn't put in." For each winner, there are millions of also-rans. The big winners may want to see their children succeed as they have, but the children have been dealt genetic "hands" of their own – gene combinations different than their parents – and it is unlikely that they will succeed as well. This is the same process that occurs in natural populations, the difference being that winning or losing in the competition for survival means living and having offspring or dying and being a genetic dead end.

CHAPTER 6

Common Descent

Dinosaurs grew fat and lazy on the luxurious vegetation in the Garden of Eden 6,000 years ago. Unfortunately for them, when Adam and Eve ate the forbidden fruit and were expelled from the garden, the dinosaurs also lost their innocence and some were then forced to feed on other creatures. Thus it was that *Tyrannosaurus rex* and the other ferocious meat eaters came into being. About 4500 years later the dinosaurs were swept from the face of the Earth by the Great Flood save for pairs of baby dinosaurs which Noah had taken aboard the Ark. When the Ark landed and spilled its cargo out onto the now dry earth, the babies grew to adulthood, mated, and repopulated. The bones of their parents meanwhile, littered the earth and formed the fossils we find today.

An interesting if somewhat fanciful story, the kind parents make up to tell their small children at bedtime. Stories like this show the lengths to which some fundamentalists Christians will go to justify the story of creation in Genesis. It is a mockery of intelligent thinking. Even children know the difference between fairy tales and reality; and yet all across the United States today in theme parks, church schools, and churches preposterous stories like this are being told to justify the creation story.

The problem with stories like this is that there is not a shred of evidence to back them up. They are based upon wishful thinking to a degree that is exceptional.

Compressing the geological and biological history of the Earth into 6,000 years is like a five second documentary detailing all of human history.

The Earth is billions of years old, not a few thousand and evolution is a fact: living things do change with time in response to environmental change. Natural Selection is a theory of a mechanism of evolutionary change which is accepted overwhelmingly by the scientific community. It has withstood the test of nearly 150 years of exhaustive research and critical study. Biologists believe that life originated only once and that all forms of life since then descended from the original form. The most compelling evidence for this theory is the remarkable similarity at the cellular level of all forms of life from bacteria to humans. The universality of the DNA code is, by itself, much too complicated to have arisen independently multiple times.

Biologists believe that life originated spontaneously under the conditions that existed on the early Earth. They are no more justified in this belief than theologians who argue that God created the first life forms directly. Both are leaps of faith, and neither can stand up to scientific investigation and experimentation. Unless scientists can show that life continues to form today from inanimate matter through natural processes, the question of how life originated on Earth will forever remain scientifically unanswered.

Speciation

Speciation is the formation of new species from preexisting ones. Currently, there are several models by which new species are thought to arise. One involves a change through time. If one could follow a terrestrial species through 100 million years of evolution and compare the living members of the species with those of 100 million years ago, they might be so significantly different as to be unable to recognize each other and breed together. Still, they are linked in the same manner that an adult is linked to himself as a child. A species changing through time, however, does not explain how one species can become two, or ten, or thousands.

The process by which new species arise from preexisting ones is incompletely understood today. One model of speciation which is widely accepted is that of geographical separation of a species. An example of this is the unique group of animals Darwin found on the Galapagos Islands. Separated from their parental stock and faced with an environment much different from the one in which they evolved, natural selection quickly changed them to such an extent that they could no longer interbreed with members of the original stock. By definition then, they are a new species.

A case in point would be the Galapagos sea iguanas. They differ from iguanas in South America by being completely black and swimming in the cold ocean in order to collect and eat algae growing there. The Galapagos are not very old as land masses go – about five million years – and they are for the most part volcanic deserts. Where did the Galapagos lizards come from? The nearest land with iguanas is South America, 500 miles away. Iguanas are cold blooded animals, and they can go for months without food. Obviously the original iguanas didn't swim to the Galapagos, but great rafts of trees torn loose by storms in the forests of South America are often seen floating down rivers and out to sea. Should iguanas be trapped on one of these rafts the leaves of the trees would provide food and water, at least for a time, and if eventually the trees floated up onto one of the Galapagos Islands the lizards would have arrived, but how could they have survived in an environment so different from the forests they were adapted to?

Surviving on the hot, black rocky shores of the Galapagos eating seaweed and whatever else they could find would be extremely difficult for animals accustomed to tropical and semitropical forests, but having left all of their predators and competitors behind they apparently were able to do so. Once established, natural selection would take over and mold them into a lizard much better adapted to the new environment and food source. How quickly this happened or whether it happened at all would be highly dependent on the ability of the animals to find food that they could eat as well as the degree of genetic variability within the species.

The rocks along the shore are black, and swimming in the icy cold waters coming up from Antarctica would quickly drain heat out of the lizards, so black would be a more appropriate color for them than green. It would allow them to raise their body temperatures more quickly after swimming in the cold water for food, and the color would provide better protective coloration. Green iguanas develop a pattern of broad black bands on the tail, so the color transformation to all black would be a very small genetic adjustment. Given the advantages of a darker skin, the darkest individuals in each generation would be expected to live better, longer lives and therefore to contribute more offspring to the next generation. In time, all lizards on the island would be completely black.

Iguanas in Central and South America can swim perfectly well but they choose not to do so unless they are thrown or fall from a tree into the water. So, in a sense the first lizard inhabitants of the Galapagos were preadapted to their eventual life of swimming in the ocean for food. All it would take is a few animals hungry enough to risk entering the rough surf for the seaweed which they could see in the clear water. These would survive while others would suffer and perhaps die of starvation, and the behavior would be perpetuated.

Green iguanas will eat a wide variety of foods. I have kept one in my laboratory for nearly twenty years. During that time it has eaten a wide variety of fresh vegetables and fruits, Cheerios, leftover pasta from the dinner table, monkey biscuits, dandelion leaves, mulberry leaves, pumpkin bread, dry cat food and a variety of other things. So, a hungry green iguana eating seaweed is not a stretch.

Over time the complete isolation of the Galapagos iguanas from the South American stock would cause them to develop into a separate species as evidenced by the fact that members from the two groups would no longer recognize each other as potential mates, or if they did, the hybrid offspring would be sterile or not survive.

There are two different species of iguanas on the Galapagos. The second species are desert iguanas that live in the highlands rather than on the seashore. This indicates that a second later invasion of the islands took place, probably by the same process that brought the first group. Which came first is not known, but the two groups evolved very differently to make use of the resources available to them. Let's assume that those that became black and lived on the seashore arrived first and adapted to a life by the sea. The second group to arrive at a much later time would not have the choice of living by the sea and feeding on seaweed because they could never out-compete those that arrived first and were now well adapted to that way of life. Chances are good that the new arrivals would be driven away from the shore by hunger and the iguanas already living there. If they were to survive, they would have to use resources and space not already taken. Apparently they were able to do so and thus could coexist with the first arrivals by living in the uplands and feeding on the vegetation there. In this way one species of tropical iguana apparently gave rise to two new species of Galapagos iguanas.

Three hundred million years ago all of the continents came together to form the super continent Pangaea. Pangaea broke up about 180 million years ago and the continents as we know them today began to move apart like giant Noah's Arks, each carrying the original plants and animals of the super continent. Once separated the species on each continent began to change until each continent developed the unique species that we find on them today. This is geographical speciation. The breaking up of Pangaea with its cosmopolitan flora and fauna led to an increase in the number of species world-wide. Should the continents join together in the future, the number of species on the planet will decrease as plants and animals from the various continents compete for resources.

Another model of speciation that is known to occur only in plants is polyploidy, the doubling, tripling, or quadrupling of chromosomes. The process by which this occurs is not very well understood, but it can happen during reproduction, and when it does, the offspring are an instant new species which cannot reproduce with the parental stock. If this happened in an animal the creature would live out its life without ever leaving offspring because it would have nothing to mate with but most plants can reproduce vegetatively by sending up new plants from the roots. These can then interbreed with each other to produce a larger population. Polyploidy can produce a new species in one generation. Most of the plants we use for food are polyploids – coffee, oats, peanuts, tobacco, apples, wheat, and sugar cane are examples. While 30-70 percent of today's flowering plants arose through polyploidy, animal polyploids are extremely rare. Polyploidy is the only known example of "instant" speciation.

Evolution by "Jerks"

Current evolutionary theory does not account well for the sudden appearance in the fossil record of new groups of plants and animals. Charles Darwin, with some frustration, referred to the apparently sudden appearance of the flowering plants and their rapid spread over the earth as "that damned mystery." According to Darwin's theory of natural selection, species change slowly over long periods of time and there should be transitional forms when one group gives rise to another. Transitional forms are uncommon in the fossil record. A few, like Archaeopteryx, a fossil first found in a limestone pit in Bavaria in 1855, seem to be transitional forms, but these are rare. The 1855 fossil was examined and originally classified as a small dinosaur. In 1861 a fossil was found with the impression of a single feather which is identical to the feathers of modern birds, clearly showing that Archaeopteryx was a bird, not a dinosaur. Several later fossil finds of Archaeopteryx also show unmistakable impressions of feathers. Archaeopteryx had a long bony tail with feathers projecting from each side, teeth in both the upper and lower jaws, and claws on its wings, all characteristics of dinosaurs but not of birds. The feet were similar to those of modern birds rather than of dinosaurs. The mixture of reptilian and avian characteristics in Archaeopteryx seems to point to dinosaurs as the parents of modern birds and makes Archaeopteryx a transitional form.

The argument is sometimes made by those who dispute evolutionary theory that *Archoeopteryx* cannot be a transitional fossil between birds and dinosaurs **because it is a bird**. This ignores the fact that despite its bird appearance *Archoeopteryx* shares more characteristics with dinosaurs than with modern birds. No other living animal has feathers, but distinct impressions of feathers have been found with dinosaur fossils found in China, and hollow bones are characteristic of both birds and dinosaurs. Birds, on the other hand, have characteristics which were not found in dinosaurs such as an opposable toe. Modern birds have three toes forward and one back. This allows them to cling to a branch. Dinosaurs did not have an opposable toe.

Children of God: Children of Earth

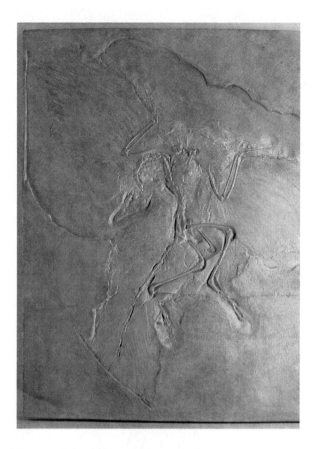

Archaeopteryx lithographica. From original fossil 150 million years old.

An explanation for the apparent rarity of transitional fossils was put forward in 1972 by Niles Eldridge and Stephen Jay Gould. They pointed out that the formation of new species is most likely to occur at the periphery of large populations where conditions are the most unsuitable for the species and where small groups of individuals are more likely to become isolated from the larger body. They also pointed out that small groups from a larger population often have gene frequencies that are not representative of the larger population.

When part of a group is separated from a larger population some genes may be lost while others may be present in unusually large frequencies. In other words, when a small number of individuals is chosen at random from a large population, the smaller group will not only lack some of the genes present in the larger population but the proportion or frequencies of its genes will also differ from the larger population.

James R. Curry

Archaeopteryx reconstruction by Greg Septon. Reprinted by permission of the Milwaukee Public Museum

If an alien spaceship landed on earth, for example, and picked up a hundred people at random, regardless of how racially diverse the group was they could not possibly be carrying all of the genes represented in the entire human gene pool. Not only that, but the genes they were carrying almost certainly be present in different frequencies from those of the general population. Consequently, if they were put onto a distant planet and allowed to populate it, they would immediately begin to evolve in ways different from those on the home planet, and given time, would presumably become a different species.

The combination of geographical isolation, atypical gene frequencies, and intense natural selection for change will, according to this concept, quickly (hundreds or thousands of years) lead the small group to become a new species. Since the transition occurs rapidly in small populations, the probability of finding a transitional fossil is very remote. Once formed, the new species may increase its numbers and enjoy a relatively long period of stability (equilibrium) before the process is repeated. The concept is called punctuated equilibrium but has been dubbed "evolution by jerks" by its detractors. Punctuated equilibrium does not contradict Darwin's theory. The process still occurs gradually and with transitional forms.

An Explosion of New Life Forms

There are five recognizable mass extinctions in the fossil record. During each a minimum of over 50% of the species on the planet disappeared. In one, the Permian-Triassic transition, 95% of all marine organisms and 70% of all land plants and animals disappeared. What is incredible about these events is the extraordinary creative impulse that followed each of them. The extinction of the dinosaurs, for example, was followed by an explosion of new mammalian and avian species which filled the vacant ecological niches left by the dinosaurs.

Nature seems to take a long time to perfect a prototype plant or animal, but once it is perfected it can radiate into thousands or tens of thousands of different species rather quickly (in geological terms). This is not too different from what happens with human technology. The development of a workable steam engine, airplane, automobile, or computer took decades, centuries, or even millennia to develop. Once a working prototype was developed, however, the technology was applied very quickly into a multitude of different forms.

The development of the automobile will illustrate. Two and four wheel vehicles drawn by animals were common for thousands of years before the first modern automobile was built. Efforts were made to develop a self-propelled or auto-mobile vehicle as early as 1335 when Guido da Vigeuano built a windmill-like vehicle with gears which turned the wheels. Leonardo da Vinci designed a tricycle driven by a clockwork device but it was never built. In 1678 Ferdinand Verbiest designed and had built a steam powered vehicle which weighed 8,000 pounds and moved at the dizzy speed of 2mph. Etienne Lenoir patented the first practical gasoline engine in 1860, but the fuel-air mixture was compressed in a separate device rather than in the cylinder. The first prototype of a modern gasoline engine was built by Gottlieb Daimler in 1885, and the first practical automobile powered by an internal combustion engine was constructed by Karl Benz in 1886.

The point is that the development of the first prototype of a modern car took thousands of years to develop and occurred in a step-wise manner. Perhaps the earliest invention which eventually led to a workable car was that of smelting metals. This was followed with the invention of the wheel. Carts, wagons, carriages and an assortment of other animal-drawn vehicles were then developed, but without some sort of motive power, an auto-mobile was still impossible. Steam engines were tried but they were too heavy to travel on roads and city streets and were relegated to iron rails. Various fuels were tried before gasoline was available, but none worked well. Braking methods were taken from the horse-drawn vehicles of the day – wooden blocks pressed against the iron rims of the wheels.

The development of casting methods which could produce the parts necessary to build an internal combustion engine led to Daimler's engine. Early cars were driven with a chain like a bicycle and didn't look much different from carriages. Because horses pulled from the front, engines were also put in the front of cars. Many different methods of motive power were tried, and by a process of selection the ones that didn't work well were abandoned and those that did work were further refined. Karl Benz and Gottlieb Daimler finally put all of the workable ideas together and the first modern car was built.

Once the long period of development was over an explosion of different vehicle types was created in an extraordinarily short period of time, all of which followed the basic workable design of the original prototype. Within a few decades trucks of all sizes, buses, hundreds of different kinds of motor cars, motorcycles, military vehicles, steam rollers, taxi's, etc. were created, all of which were based upon the original Daimler-Benz concept. Henry Ford built and sold his first car in 1903 and two decades later 15 million model T's were on the road.

The development of airplanes followed a similar pattern to the automobile. People had been trying one device or another to get off the ground for thousands of years, but after the Wright brothers flew the first successful prototype in 1903 their basic design was quickly turned into dozens of different airplanes, each with a specialized function. Less than 40 years after their first flight at Kitty Hawk jet planes were carrying people around the globe at nearly supersonic speed.

The process of invention and development in living organisms is in many ways similar to the development of the automobile or airplane. It may take thousands, hundreds of thousands, or millions of years to develop the prototype for a new kind of plant or animal. As a result of random mutations, thousands of different ideas are tested in the crucible of natural selection:

those that work are further refined; those that don't are discarded. Many more are discarded than survive. Mutations that produce new materials are especially important. A reference to the importance of new materials in the development of human technology will illustrate the point.

Each time a new material has been introduced into human culture, new possibilities and opportunities have opened up. Stone and wood were the first materials used and they radically altered the human condition. A carcass that could never have been broken into with fingernails and teeth could be sliced up with a sharpened stone in a matter of minutes. Eventually, the ancients became so proficient in the use of stone that great buildings like the pyramids were constructed.

Learning to make and use metals was another giant leap forward. At first copper, then bronze, then iron, then steel - each time opening new technological possibilities. Skyscrapers could never have been built with the earlier materials, but the ability to produce high grade steel made them possible. In more modern times, the use of titanium and carbon composites have made possible aircraft that can fly at supersonic speeds, something that never could have been done with iron or steel. Concrete, another innovation in materials has made modern cities, roads, and bridges possible.

The same principle holds in living things - produce a new structural protein and make something possible that was not possible before - an insect wing, a birds' beak, a sharks' skeleton, an elephant's tusk, a reef fish's brilliant coloration. The materials from which all of these are made came originally from chance mutations. Changes in genes which affect physiology provide similar opportunities for advancement. A change in the chemical structure of an animal's oxygen carrying pigment may make the pigment more efficient and thereby promote survival over others without the new pigment, or changes in one of the chemicals a plant produces to protect itself from being eaten by herbivores may make it immune to predation.

The development of the first prototype of a modern bird took millions of years. Before a modern bird could emerge feathers had to be developed from a very strong, lightweight, wear-resistant material. Keratin, a protein, was ideal, and it was transformed into a variety of feather types, each with a specialized function: flight, insulation, color, streamlining, etc. Keratin was also suitable for use in a beak, so the teeth were eliminated and the head made lighter and more streamlined. Once the prototype of a modern bird was developed all that was needed was an opportunity for them to succeed. This came with the demise of the dinosaurs and the opening of ecological niches formerly dominated by those reptiles. The birds exploded into those niches, producing tens of thousands of species from the original prototype, and as with automobiles, all of the species have the same basic characteristics as the prototype. The process of developing many species from one is referred to as radiation. As in the development of the automobile or airplane, the process of evolution does not take a straight line from the starting point to the end point. Many blind alleys are followed to their dead end in both cases.

There is a critical difference between human technological development and evolutionary development. Human development has a purpose, a goal, and movement is always toward that goal. Evolution has no purpose or goal except survival in the present and successful reproduction. Natural selection weeds out the successful from the unsuccessful in each generation. There is no forethought or purpose in the process. Many people would like to believe that humans were the ultimate goal of evolution. From everything we know about the processes and mechanisms of evolution that is impossible.

This should not be taken to mean that the process of evolution is random. Clearly there is an organizing principle at work. That principle is natural selection. It can turn a dinosaurs foreleg into a bird's wing, but only if the transformation has a selective advantage in real time. But what would the advantage be? A primitive wing could be used for gliding from tree to tree or from tree to the ground. This would aid in escaping predators and is an efficient method of moving from place to place. Beating the wings while running also increases speed, as anyone knows who has ever chased a chicken around the barnyard.

Initially, then, forearms with feathers on them are an aid to survival and would be selected for. In each following generation those individuals with slight improvements on the primitive wings which helped them soar better or run faster would have an advantage that enabled them to live longer and leave more offspring. Step by step the process would lead to the development of a modern bird, but without any forethought or goal in that direction. The proof of this is in the conservative nature of the process. Everything new, like a flying dinosaur (bird), is developed by modifying what already exists. At no point in evolutionary history are new plants and animals created out of whole cloth.

Some theists, cosmologists, and biologists have postulated that given time, more intelligent forms of life will evolve. This assumes that evolution is directional and purposeful. New forms of life will certainly emerge with time, but they may be less intelligent, not more. Evolution is not about intelligence - it is about using resources efficiently, surviving long enough to leave large numbers of offspring, avoiding being eaten or killed by predators, surviving the attacks of pathogens and parasites, adapting to climate change – it is about survival. Quite frankly, our species is not the leader in the race for survival. Personally, if I had to bet on which animals have the best chance of surviving and flourishing long into the future, it would be those beautiful little animals very few people hesitate to step on or swat: insects.

Biologists don't understand all of the forces that drive evolution any more than geologists understood the forces of plate tectonics a few decades ago, but the search goes on for answers. Without question, exciting revelations lie ahead which will help us to better understand the natural world and accept our place in it.

> *The more I look at the great complex of the animal world, the more sure do I feel that we have not yet reached its hidden meaning, and the more do I regret that the young and ardent spirits of our day give themselves to speculation rather than to close and accurate investigation.*
> - Louis Agassiz

CHAPTER 7

Why Males?

A Woman without a man is like a fish without a bicycle.
From a bumper snicker.

In her dark retreat in the ground the female sat and brooded over her fate. Something within was making her nervous and irritable, and it was growing worse by the day. The walls of the tunnel were lined with silk, but the dampness of the soil penetrated and added to her anxiety. From the mouth of the tunnel a white silken web spread out funnel-like into the vegetation. Thin guy ropes stretched from the edges of the web to stalks of vegetation which kept the web taut.

Suddenly, she felt the web vibrating, as if an intruder were walking across it. She tensed, and prepared to attack. The vibrations quickly increased in intensity and frequency, causing her to move cautiously toward the tunnel entrance. As she moved, an inexplicable wave of excitement began to build in her. The pulsing of the web surrounded her with vibrations, and despite her efforts to resist, she began to relax. A dream-like feeling of well-being spread through her and she submitted to the magic enchantment of the vibrations. At the same time, an irresistible passion exploded within, causing her to tremble with suppressed excitement. The excitement continued to build until it was almost unbearable, and begged for release.

A small figure approached her carefully. Gently, ever so gently, his legs caressed her and the trembling within was calmed. She resisted the impulse no longer, but gave herself to his advances. Without hurry, he continued to stroke and reassure her. At the appropriate moment, he inseminated onto the web and dipped his pedipalps into the semen. When they were full, he thrust them into her genital opening. After several thrusts, he disengaged her, and cautiously backed away.

For a long moment she rested, feeling relaxed and satisfied. A drama hundreds of millions of years old had been acted out with the precision of a ballet. The tiny male spider was fortunate to have escaped with his life. Having performed the only function for which he exists, he could have served as a meal for the female and provided much of the protein necessary for her to produce their offspring.

Sexual reproduction is a critical part of the life cycle of every organism, but sex is an extremely inefficient and demanding way of reproducing. Male animals must spend time searching for a potential mate, compete with other males when one is found, and court females. All of this involves an enormous amount of time and energy which could instead go directly into producing more offspring. Plants also expend an enormous amount of energy producing pollen or other male gametes and then finding ways to ensure that a very few reach the ovary of female plants and produce seeds. This often involves forming evolutionary alliances with animal pollinators like insects, birds, bats, or mammals, but these are expensive because the pollinators must be rewarded with pollen and nectar for their work. Animals pollinate more than ninety percent of all the flowering plants in the world.

These are not the only disadvantages to sexual reproduction. Organisms which reproduce without sex (asexually) can reproduce more rapidly, and they pass on all of their genes to their offspring. The offspring of sexually reproducing organisms get a random mixture of genes from both parents, which can dilute and destroy the successful combinations of genes in each parent. Consequently, many of the genetic traits which allowed the parent to be successful may be diluted or lacking in the offspring.

An example of the extraordinary costs of sexual reproduction is the reproductive migration of European eels. Adult eels from all over the continent begin leaving their ponds and streams in July and move toward the sea. Their destination is the Sargasso Sea, an oval-shaped area thousands of square miles in size north of the equator and south of Bermuda. The Sargasso Sea is very calm, with vast amounts of floating seaweed, and the entire sea slowly rotates clockwise.

When the eels reach the Atlantic Ocean the transition from fresh to salt water requires significant changes in the structure and physiology of their body, including changes in the gut which make it impossible to eat. Once they have completed the transition they begin the 4,000 mile journey to the Sargasso Sea, living on food stored in their bodies for energy to make the trip. Upon reaching the spawning grounds, they mate and lay eggs. The larvae which hatch from the eggs travel with the Gulf Stream across the Atlantic and after three years reach Europe and make their way to the headwaters and ponds of their parents. They live there for 10-14 years before beginning the return journey to their spawning grounds in the Sargasso Sea.

The incredible demands and dangers of these migrations point out how critical it is for populations of organisms to maintain a high level of genetic diversity within their gene pool. By gathering together in one place hundreds of thousands of eels from all over Europe are brought into close contact and can mate with each other. Their offspring swim away carrying new combinations of genes which have resulted from this gigantic stirring of the species gene pool. This helps to maintain the integrity of the species across a broad geographical range, and expedites the rapid spread of useful mutations. The result is a population with a highly variable but cosmopolitan gene pool which enables the species to adapt to changes in its environment. It is comparable to putting all the cards together following the completion of a hand of poker, shuffling well, and then dealing out new hands, some of which are worth nothing, and some of which can make one wealthy.

Given the extraordinary demands of sexual reproduction, why do all living organisms, even those that normally engage in asexual reproduction, reproduce sexually at least part of the time? Another way of asking the question is: What are the evolutionary benefits of sexual reproduction? Evolutionary theory holds that natural selection favors traits that increase the probability of survival and eliminates traits that decrease the chances of survival. Evolutionary biologists believe that the advantages of sexual reproduction override the substantial disadvantages in several critical areas:

Adapting to a changing environment.

The spread of favorable mutations.

Diluting the effect of harmful mutations.

The Red Queen

In Through the Looking Glass, Lewis Carroll's 1872 sequel to Alice's Adventures in Wonderland, Alice finds herself a pawn in a giant chess game. At one point she complains to the queen that she doesn't seem to be getting anywhere even though she is running as hard as she can. The queen admonishes her that in the looking glass world constant exertion is required just to remain in place.

Alice never could quite make out, in thinking it over afterwards, how it was that they began: all she remembers is, that they were running hand in hand, and the Queen went so fast that it was all she could do to keep up with her: and still the Queen kept crying 'Faster! Faster!' but Alice felt she COULD NOT go faster, though she had not breath left to say so.

The most curious part of the thing was, that the trees and the other things round them never changed their places at all: however fast they went, they never seemed to pass anything. 'I wonder if all the things move along with us?' thought poor puzzled Alice. And the Queen seemed to guess her thoughts, for she cried, 'Faster! Don't try to talk!'

Not that Alice had any idea of doing THAT. She felt as if she would never be able to talk again, she was getting so much out of breath: and still the Queen cried 'Faster! Faster!' and dragged her along. 'Are we nearly there?' Alice managed to pant out at last.

'Nearly there!' the Queen repeated. 'Why, we passed it ten minutes ago! Faster!' And they ran on for a time in silence, with the wind whistling in Alice's ears, and almost blowing her hair off her head, she fancied.

'Now! Now!' cried the Queen. 'Faster! Faster!' And they went so fast that at last they seemed to skim through the air, hardly touching the ground with their feet, till suddenly, just as Alice was getting quite exhausted, they stopped, and she found herself sitting on the ground, breathless and giddy.

The Queen propped her up against a tree, and said kindly, 'You may rest a little now.' Alice looked round her in great surprise. 'Why, I do believe we've been under this tree the whole time! Everything's just as it was!' 'Of course it is,' said the Queen, 'what would you have it?' 'Well, in OUR country,' said Alice, still panting a little, 'you'd generally get to somewhere else--if you ran very fast for a long time, as we've been doing.' 'A slow sort of country!' said the Queen. 'Now, HERE, you see, it takes all the running YOU can do, to keep in the same place. If you want to get somewhere else, you must run at least twice as fast as that!'

Alice surrounded by the characters of Wonderland in The Nursery "Alice" (1890)

Biologist Leigh Van Valen of the University of Chicago has used the theme of the Red Queen to point out that in nature species must constantly evolve to maintain their place in the hierarchy of life. Change is not only good, it is essential, because extinction is the fate of those organisms which fall behind in the race to survive. Change is possible only through sexual reproduction, and the associated costs are the price every species must pay to stay alive. Parasites like bacteria, viruses, fungi, and worms mutate frequently, producing new combinations of genes that can overcome the defenses of potential plant and animal hosts. Unless the hosts can counter with genetic changes of their own, they are likely to fall into the dark pit of extinction. Likewise, an arms race goes on constantly between predator and prey and between competitors for the same resources.

The game of survival among living organisms is like the competition which takes place among football teams: teams which never change their offensive and defensive strategies sooner or later become losers. A team which can come up with a powerful new offense will take its rivals by surprise and be extremely successful for a few years. Gradually, however, other teams will learn to defend against the new strategy and may get so good at it that they completely shut it down. This requires new thinking on the part of the offense, which in turn requires adjustment by the defense: the adjustments never end, and the game is constantly changing. The same kind of "coevolution" occurs in living organisms. A plant develops a new defensive chemical to ward off insects that prey on it: the insects, which are probably highly dependent upon the plant for their survival, must now respond or lose their main food source. They may respond so well that they begin to destroy all of the plants. It is now up to the plant to again make adjustments which will bring the insects back under control. Like the football strategies, it goes on relentlessly. When one side or the other fails in this competition to survive, it becomes extinct. This is the Red Queen Hypothesis: to remain in place, you must run constantly.

It is estimated that ninety percent of the indigenous human population of North America were killed by ordinary European diseases like measles and small pox because the diseases were new to them and they had no resistance. The population survived because some members of the population did have resistant genes to the new diseases, despite never having been exposed to them. Since the discovery of inoculation, antibiotics, and other medical "miracles" our adjustments to the rapidly mutating microorganisms and viruses has been technological rather than genetic, but we are discovering that the Red Queen Hypothesis also applies to our technology. When penicillin and sulfa drugs were discovered in the late 1930's they were true miracle drugs. People in their death throes from bacterial pneumonia, scarlet fever, whooping cough, and a host of other bacterial diseases were brought back to life and health with a speed that amazed the doctors attending them. They were true "miracle" drugs, but within a few short years it became apparent that the bacteria were mutating to resistant forms. We have been able to keep ahead of the pathogens by constantly developing new antibiotics and strengthening the older ones, but people in the medical profession are seriously worried about the increase in the number of bacterial pathogens that no longer respond to our antibiotics.

Bacteria are not about to succumb to the threat of antibacterial products like antibiotics. Using these products unnecessarily is causing the bacteria to mutate to resistant forms which many health experts fear is bringing on a global health crisis. The bacteria have been practicing the Red Queen Hypothesis for several billion years and understand it extremely well.

Asexual reproduction is essentially a process of cloning, meaning that all of the offspring are genetically identical. Given that natural selection and evolution rely on an enormous amount of genetic variation within populations, organisms that lack variation can never adapt to a changing environment or colonize a new habitat. Every organism is constantly faced by changes in its physical and biological environment. If all members of a population are identical genetically they will all survive these changes or they will all die without reproducing

Mutations are random changes which occur in the genes during the production of eggs and sperm. Because they alter successful genes, most mutations are harmful and natural selection keeps them at low frequencies in the population. Asexual organisms can accumulate more and more harmful mutations with time. Eventually, if the burden of mutations becomes too great, the organism will succumb to extinction. Sexually reproducing organisms can dilute the effect of harmful mutations through the combining of male and female genes in the offspring.

For a harmful mutation to show up in sexually reproducing offspring a copy of the mutant gene must pass to them from both parents, and since harmful mutants are kept in very low frequency within the population by natural selection the probability of this happening is very low. An example is the cystic fibrosis gene in humans. This recessive gene is a mutant form of a normal gene. The effect of the mutant gene is a chronic disease that causes the body to produce unusually thick, sticky mucus that clogs the lungs and can lead to life threatening infections. The gene also prevents the pancreas from producing the enzymes which digest food. Until the middle of the last century children with the disease rarely lived to attend elementary school. The desease occurs with a frequency of 1 in 2,500 Caucasian people in the U.S. and in lower frequencies among other nationality groups. Even when both parents carry the mutant gene the chance of having an affected child is only 1 in four.

Mutant genes are usually recessive to their normal counterparts; consequently, individuals carrying only one gene for the mutation will not be affected by it. Individuals carrying two copies will die prior to reproducing or be less successful at surviving and reproducing. Each time death occurs two of the mutant genes are eliminated and the frequency of the genes in the population is kept low.

It is nearly impossible to completely remove even the most harmful mutant gene from a population because individuals carrying one mutant and one normal gene are not harmed by it, and because normal genes mutate spontaneously from the effects of ultraviolet radiation; thus new mutants are constantly forming. It is estimated that every human carries from one to three mutant genes which were not present in their parents.

Mutants are often thought of as mistakes. It is better to think of them as random changes which do not work well for the organism in the present but which may have value in the future. Remember that all innovation in evolution comes from mutations, many or most of which were harmful initially. Without mutations, evolution would grind to a halt.

The mixing and recombining of genes that occur in sexual reproduction allows favorable mutations to spread through a population much more rapidly than in an asexual population, thus greatly speeding up the process of evolution. This is critical because most sexually reproducing organisms have much longer life spans than the asexual organisms which attack them. A long lifespan generally means a greater span between generations, and this greatly reduces the speed with which the species can undergo adaptive change. A gene mutation which occurs in an asexually reproducing organism will be replicated and passed to its offspring during reproduction, but it will not be shared with other lines of organisms which are cloning themselves.

Having pointed out what appear to be advantages to sexual reproduction, it must be admitted that sex works against individual organisms in their attempts to pass their genes into the future. Even the most highly successful individual plant or animal can pass on only 50% of its genes, and these must be mixed with 50% of the genes of another individual. This makes for a real crap shoot. There are benefits to the species as a whole, but individuals lose out. Only rarely in the human population do great musicians, artists, scientists, athletes, etc. have offspring as capable as themselves, but in each generation exceptional people do surface, often from parents who lacked their children's natural talent.

Asexual reproduction persists in many plants and microorganisms because of the speed with which populations can increase in size. A single bacterial or fungal spore landing on a piece of food, for example, can in a matter of hours increase into millions to take advantage of the food source before competitors arrive. Likewise, a rapid increase in asexually reproducing pathogenic organisms allows them to overwhelm the host before its defense mechanisms can come into play. Organisms like bacteria and others that normally reproduce asexually have the ability to engage in sex by joining together and exchanging chromosomes through a bridge of cytoplasm. This is not reproduction, but it is sex because it creates new combinations of genes and allows adaptation and evolution to occur.

So God created man in his own image, in the image of God created he him; male and female created he them. Genesis 1: 27.

None of the benefits of sexual reproduction completely explain why mating types (males and females) should exist. Most flowering plants contain both anthers, which produce pollen, and stigmas, which produce eggs, but a number of devices exist to prevent self-pollination because this would be a severe form of inbreeding, which is counter to the purpose of sex. Earthworms, slugs and a number of other animals contain both sexes in each individual. Called hermaphrodites, any two individuals can meet and exchange gametes. This would seem to be a far more efficient and rapid method of producing offspring: the competition that goes on among males of other species is not necessary, the time and energy it takes to find and court a mate is greatly reduced, and both individuals go away from the encounter to produce offspring. Still, hermaphrodites make up only a small percentage of all animal species.

Sex Run Amok

"The sight of a feather in a peacock's tail, whenever I gaze at it, makes me sick!"
Charles Darwin, in a letter to botanist Asa Gray, April 3, 1860

The male peacock's magnificent tail, a moose's gigantic rack, a bower bird's large collection of colorful objects, the head butting of mountain goats – all of these male courtship structures and activities require enormous expenditures of food energy to produce and maintain, and because they attract attention, they increase the animal's vulnerability to predation. Why then does natural selection not eliminate such ornate but useless structures and behaviors? Charles Darwin saw the problem as a challenge to his theory of natural selection and worried over it, as the above comment to Asa Gray attests. In his last book, The Descent of Man and Selection in Relation to Sex, Darwin proposed that ornaments and behaviors of this kind are selected for by females during courtship, and that sexual selection by females is one of the great driving forces of evolution. Since males are somewhat expendable following insemination of females (in quite a few species males are devoured following copulation by females hungry for protein to put into their eggs), and since one male can inseminate many females, the development of secondary sexual characteristics which will entice females to mate makes evolutionary sense.

It is a general rule that in courtship behavior among animals, males display and females choose. If females are choosing mates on the basis of brilliant colors or huge antlers, then colors will get brighter and antlers will get larger as the generations pass. Naturally, there is an upper limit to how outlandish secondary sexual characteristics can get. If an elks' antlers get so large that it can no longer hold its head up or run, natural selection will quickly take care of the problem by feeding the animal to the wolves.

Children of God: Children of Earth

PeacockDisplay
By David Wagner

He approaches her, trailing his whole fortune,
Perfectly cocksure, and suddenly spreads
The huge fan of his tail for her amazement.

Each turquoise and purple, black-horned, walleyed quill
Comes quivering forward, an amphitheatric shell
For his most fortunate audience: her alone.

He plumes himself. He shakes his brassily gold
Wings and rump in a dance, lifting his claws
Stiff-legged under the great bulge of his breast.

And she strolls calmly away, pecking and pausing,
Not watching him, astonished to discover
All these seeds spread just for her in the dirt.

Reprinted from "Best of Prairie Schooner: Fiction and Poetry," University of Nebraska Press, 2001.

Sperm are, to coin a phrase, a dime a dozen. Actually a dime for millions would be more accurate. Compare the small half ounce or so of semen which a rooster provides during copulation with the eggs produced by the hen. The hen must then incubate the eggs for 21 days and protect the chicks until they can take care of themselves. The male's contribution was a few seconds of exertion during copulation and a penny's worth of semen. Since it is the goal of both male and female to extend their genes in time and space, and given the large discrepancy in what the male and female bring to the reproductive table, the two sexes obviously require different strategies for achieving their respective goals.

For males, the best chance for reproductive immortality is by inseminating as many females as possible. This, of course, brings males into competition with each other, and fighting and killing can result, although in many species the competition for mates is through ritualized fights and displays. David Wagoner's vivid description of a male peacock courting a female illustrates the generalization that males display and females choose. Strutting males and indifferent females is a scene common to many animal species, including our own. In many species the only contribution that the male makes to reproduction are the sperm cells which fertilize the female's eggs.

Females, on the other hand, have a much greater investment in the production of offspring and are likely to be looking for quality rather than quantity in a mate. Female red-winged blackbirds, for example, cruise through the territories of male birds and choose a mate based not upon his qualifications, but upon the quality of his territory. Studies have shown that females will choose to be the second or third mate of a male on a good territory rather than the only mate of a male on a poor territory, even though they will get less help from the male in feeding and protecting the young. A good territory is one with excellent nest sites to protect their eggs and young and abundant food to nourish them.

A female judging the quality of a male's territory is really attempting to give her offspring the best possible chance of surviving and reproducing successfully, but how does choosing a male with a huge tail or large antlers benefit a female and her offspring? A number of hypotheses have been put forth. What the female must do in selecting a mate is to gauge the *fitness* of the male to be a parent to her offspring. Perhaps the best proof of genetic fitness is old age. The longer an animal can survive the attacks of predators, disease organisms, and parasites the greater its fitness. If the exaggerated secondary sexual characteristics increase in size with age, this is a sign of superior fitness. In fact, antlers and a peacock's tail are larger in healthy older males than in younger ones.

Truth in Advertising

Producing supernormal structures requires a large expenditure of energy. Animals having difficulty feeding themselves, sick animals, and those riddled with disease will be unable to expend such energy; thus they cannot fake fitness they do not possess. An Israeli biologist, Amotz Zahavi, has proposed what he calls the **Handicap Principle**. In essence, the Handicap Principle says that a moose with a large rack is advertising to females that even with the severe handicap of these gigantic antlers, I can survive. Likewise for the male peacock: I can't fly, I am brilliantly colored and can't hide, and I must find an enormous amount of food to produce these incredible feathers, but despite these handicaps, I am prospering. In other words, I have superior genes and am worthy to father your offspring. These are things which cannot be faked by unhealthy or otherwise unfit males.

In order to ensure that their genes, and not the genes of another male, pass into the next generation, some animals take extraordinary measures.

Some male insects like dragonflies have a penis which can remove the sperm of a previous mating and replace it with their own. They must then guard the female while she lays eggs to prevent another male from replacing their sperm. Some male slugs and snakes produce semen which coagulates in the female producing a plug which prevents another male's semen from entering the females' reproductive tract. When a young male lion displaces an older male in a pride of females, he will immediately hunt down and kill the females' cubs. This serves two purposes: it eliminates his having to protect the offspring produced by another male (there is no evolutionary gain for him in protecting another male's offspring), and it brings the females into heat so that he may mate with them and produce his own offspring. Such behavior may sound horrific to us, but consider that studies have conclusively shown that a small human child or infant is roughly a hundred times more likely to die at the hands of a stepfather or boyfriend than at the hands of a biological father.

Males and females are players in an evolutionary arms race in which both sexes attempt to manipulate each other in order to pass their genes into the next generation. Their differing strategies lead to an ongoing tension between the sexes.

The rules of the male-female game are different for each sex: the female attempts to raise the costs of sex while the male attempts to lower it.

Do these rules of sexual strategies hold for humans as well as other animals? David Buss and David Schmitt, psychologists at the University of Michigan, in collaboration with 50 other scientists surveyed the mating preferences of 10,000 people in 37 countries. They found that the process of choosing a mate is **not** highly culture bound but ….. "that human beings, like other animals, exhibit species-typical desires when it comes to the selection of a mate."

They found that for short-term mating men attempt to find women who are sexually accessible, fertile, and who will engage in sex with a minimum of commitment and investment from the man. Prostitutes fit these criteria perfectly. Women, on the other hand, attempt to identify men who have high quality genes, extract resources from them, and cultivate "back up" mates.

Their study and those of others have shown that short-term mating is more important to men than women. In one study of college students at the University of Hawaii by Russell Clark and Elaine Hatfield, students were approached by an attractive stranger of the opposite sex and asked one of three questions: "Would you go out on a date with me tonight?" "Would you go back to my apartment with me tonight?" or "Would you have sex with me tonight?" Fifty percent of the women who were approached agreed to the date, 6 percent agreed to go to the apartment, and none agreed to have sex. Among the men who were approached, 50 percent agreed to a date, 69 percent agreed to go back to the woman's apartment, and 75 percent agreed to have sex. Apparently among humans, as in other animals, the sex investing the most in offspring are also the most discriminating in choosing a mate. Women can have only one child at a time, which makes the number of partners irrelevant to them in terms of producing children, but men can parent as many children as they have sex partners.

Buss and Schmitt found that the rules change, however, for long term mating. Men face a problem not faced by women: ensuring that children are their own. When seeking a long term mate men are more concerned with fidelity than women and have used (and continue to use in some cultures like those in Muslim countries) various methods to sequester women and/or make them less desirable to other men. Buss and Schmitt found that fidelity was the characteristic most valued by men in a long-term mate, and that men are more concerned about their mates having sex with another man than they are with their mate falling in love with another man. Women on the other hand were more concerned with their mate forming an emotional attachment to another woman. Physical appearance was less important to men in choosing a long-term mate than a short-term one. For long-term mating men want women with high reproductive value (as evidenced by youth and good health), good parenting skills, and "high quality" genes.

Women want long-term mates who are willing to invest resources, provide physical protection, have good parenting skills, and have "high quality" genes. Physical attractiveness is less important to women than to men in long-term mates, but the ability and willingness to provide resources for her children is very important.

Studies have shown that people who live together prior to marriage have a higher divorce rate than those who do not. This seems contrary to common sense since people living together for months or years should have a good sense of what their partners are like prior to marriage. The problem is that people who choose to live together without marriage are applying short term criteria in their selection of a partner and are thus selecting someone who may not meet their long term criteria for a mate. When the relationship is formalized through marriage it becomes apparent to both that they chose a spouse that was fun to be with rather than one they would like to share their life with.

The work of Buss and Schmitt and others show that human mating preferences are not arbitrary or culture-bound, but universal and highly adaptive to the needs and problems faced by men and women in the course of evolution. Each of us today are the inheritors of genes passed down to us through 50,000 generations or more by men and women who were able to successfully solve the problems associated with finding high quality mates and reproducing.

Ye have heard it said by them of old time, Thou shalt not commit adultery: But I say unto you, That whosoever looketh on a woman to lust after her hath committed adultery with her already in his heart.
                              Matthew 5: 27-28

    The most difficult questions to ask are often about things we take for granted: why does an apple fall from a tree? Why does an oak tree produce other oak trees instead of mice or ferns? But questions of this type often have complex and profound answers. An obvious question for those who believe that our species was created directly by God rather than through natural processes is why did He create two sexes? Was He trying to set us up for all of the personal and social problems associated with sex? Did He not know that sex would become aberrant in the most abhorrent ways: pedophilia, rape, murder, lust, and prostitution; that it has the power to destroy people and relationships? If we are special creations of God, made little lower than the angels, why did he strap us with this most elemental and brutish form of behavior? Is it not beneath our dignity as the children of God to live lives dominated by the same primitive force that enslaves all other living creatures? Why did He give us great intelligence but tie it to an anchor that forever keeps us from the ethereal realms of pure intellect?

    From a religious standpoint sex makes no sense at all. It is a force so primal and irresistible that we have no choice but to give it a central place in our lives. It ties us to all other living creatures, no matter how primitive, in a way that makes any concept of special creation totally irrational. Sex (and men) make sense **only** when viewed in the context of evolutionary theory. Without sex we, like all of God's creatures, would eventually become extinct.

CHAPTER 8

Genesis

And God said, Let the earth bring forth grass, the herb yielding seed, and the fruit tree yielding fruit after his kind, whose seed is in itself, upon the earth: and it was so. Genesis 1: 12

Before there were human beings, long before there was this tiny blue planet we call Earth, there were dying stars whose light had pierced the darkness of deep space for billions of years. In their death throes these gigantic givers of life exploded, throwing out atoms that drifted aimlessly in interstellar space. Eventually, gravity brought some of them together into small clumps, which were attracted to larger clumps, until over untold millions of years the planet Earth formed. Gravity and the energy of radioactive materials caused the planet to heat up until it was a roaring furnace of molten materials. The heavier materials like iron settled into the interior, the lighter ones floated on top to form the crust, and the lightest of all became a primitive atmosphere of carbon dioxide, methane, water vapor and various other gases.

Some of the atoms thrown out by those exploding stars eventually became us, but first they traveled a long, uncertain path from superheated primordial ooze to the first living cells, to jellyfish, sea urchins, fish, dinosaurs, mammals, primates, anthropoids, and humans. The elements of those ancient creatures now reside in us and in this evolutionary generation of plants and animals. Molecules that were part of the hot blood of dinosaurs as they thundered across the ancient landscapes now fleetingly have their place in our own living systems. Long after we and our kind are gone from the planet the elements of our bodies will reside in plants and animals of a kind unknown to us today.

We are part of a gigantic, complex, living community that extends back in time 3.5 billion years, and that has maintained itself in the face of constantly changing environmental challenges of such magnitude that we can only stand in awe at its vitality and endurance. There have been mass extinctions, periods when as much as 90% or more of the extant life forms on Earth vanished in a short period of time, the result of environmental changes too great to respond to. But these were not failures, for each episode was followed by the most incredible explosion of new life forms – creative impulses of such magnitude as to stagger the imagination.

Someday long into the future it will all end. Our star, the sun, will die, and the Earth will no longer support life. But in its death throes the sun will scatter its atoms to drift aimlessly in space perhaps to eventually form a new blue planet where the whole process will begin again. Or, perhaps not. For all our speculation about extraterrestrial life we know for certain of only one place where life exists – the planet Earth. It may exist on countless other planets, but it almost certainly would have evolved so differently from life on Earth as to be unrecognizable to us. Until we know differently, we should assume that life on Earth is unique, a very special one time event in time and space. We should treat it with the reverence of our primitive ancestors who worshipped nature and made deities of its various forms.

We will probably never know how life originated on the Earth. Perhaps a Creator planted the seed of life and left it to develop under the direction of the special environmental conditions of this extraordinary planet. Perhaps it immigrated here from another planet, a hitch-hiker on some piece of space debris. When large objects like comets or meteors impact with a moon or planet the energy of impact is often great enough to throw chunks of debris out into the solar system. These may eventually land on other planets. Mars and Earth have been swapping rocks for millions of years, and one Martian rock found in Antarctica contains fossils that some biologists believe are bacteria-like organisms. The conditions of space are extremely severe, so it seems unlikely that life originated in this way, but a bacterial cell left on equipment on the moon was found to be alive after three years, and nobody knows how long these incredibly resistant little life forms can survive in space. Even if we were to accept that life came here from elsewhere in the universe, that still begs the question as to how and when life originated.

Most biologists believe that life originated from the combination of simple, common chemical compounds under the natural conditions which existed on the early Earth. Water is the most important of these. Earth is a water planet, as revealed from space by its beautiful blue face. The fires of life burn only in water; without it life as we know it cannot exist. Carbon dioxide spewing from ancient volcanoes could have supplied the carbon necessary to build the first organic compounds. The chemicals of living cells – proteins, DNA and the other nucleic acids, sugars, starches and fats; - are all built around a skeleton of carbon. Nitrogen, essential for the formation of proteins and nucleic acids could have been supplied by methane, a gas thought to have been common in the early atmosphere. All of the other elements needed to build and maintain a living cell - phosphorous, zinc, and a host of other minerals - are common everywhere on Earth's crust and in the oceans. For simple molecules to be built into more complex ones, however, energy is required and energy was abundant on the early Earth in the form of lightening, thunder, heat, and sunlight.

But simple chemicals do not create something as complicated as a living cell unless some incredibly powerful creative force is at work. Building large molecules from small ones is like building a skyscraper out of bricks – very stable small units are turned into large unstable units which can break apart and fall at any time because they oppose natural forces. If we are to believe that life originated under natural Earth conditions, then a creative life force - an organizing principle at the most basic level - must have been present.

Biologists believe that the creative impulse for life is as much a natural phenomenon as are mountains or seas. Fossils as old as 3.5 billion years have been found in the ancient rocks. The life forms of the planet have successfully survived every challenge since then, not through individual or species survival, but through reproduction and the creation of new species as older ones died out. Life is a force so primal and powerful that once set in motion it seemingly cannot be stopped except in the most extreme environmental conditions. It is quite possibly the most powerful force in the universe, for almost certainly the life force of the planet Earth is only a subset of some cosmic organizing force. Life is a violation of the chaos that pervades the universe, but once formed, it has incredible staying power.

Just what sequence of events could have led by physicochemical means to the advent of the first primitive cell? The improbabilities seem unsurmountable without the intervention of the designer....
- John M. Templeton and Robert L. Herrimann

For some people of faith, the answer is quite clear - God is the creative force of the Earth, and indeed, of the entire universe. Biologists cannot dispute this as scientists. Many biologists, in fact, are staunch believers in the presence of God the Creator. But God is supernatural, and biologists believe that the origin of life was borne of processes natural to the earth. The two beliefs are not mutually exclusive. If God created life in a direct way, then we need look no further for the solution to how life began. The problem for scientists with this solution is that it cannot be verified or falsified scientifically. This doesn't make it wrong; it just makes it unacceptable to people accustomed to looking at the Earth as part of a self-functioning universe.

In the beginning God created the heaven and the Earth. And the earth was without form, and void; and darkness was upon the face of the deep.
Genesis 1: 1, 2.

The Earth formed more than 4.5 billion years ago, a time line that geologists are fairly certain of based on the dating of ancient rocks from the Earth, moon, and extraterrestrial objects like meteorites. The earliest identifiable fossils of single celled organisms are found in rocks more than 3.5 billion years old. The time required for the formation of the first life forms then, was about a billion years. During this time the planet cooled, and water vapor in the atmosphere condensed and formed the oceans, but conditions were still very hostile to life. The atmosphere at the time was very different from that of today. Free oxygen, which is very reactive, was absent, having combined with metals like iron and a host of other elements. Free oxygen did not appear in the atmosphere until photosynthetic organisms evolved and began producing it as a byproduct of photosynthesis nearly two billion years after the first life forms appeared.

The discovery of extremophiles, microorganisms which can thrive at temperatures near or above the boiling point of water, has confirmed for biologists that life could have existed under the extreme conditions of the early Earth. The first extremophiles were discovered in the hot geysers of Yellowstone National Park. Microbes living there are capable of surviving at temperatures as high as 176F (80C). One group of extremophiles, the Archaea, closely resembles bacteria, but differ significantly enough from them to be placed in a separate group. Most Archeans are anaerobic and thus cannot live in the presence of oxygen. This and other characteristics make them prime candidates for one of the earliest forms of life.

The Archeans and other microbes can live in a variety of extreme habitats other than heated water, including solid rock and deep within the earth. Excavations from a South African mine at a depth of two miles contained living microbes. A bacterium with the scientific name *Deinococcus radiodurans*, but called Conan the Bacterium by its founder, is the most radiation-resistant organism known. Discovered in meat that had been "sterilized" by ultraviolet light, the bacterium is able to withstand a dose of radiation 3,000 times greater than would kill a human. Radiation levels reaching the early Earth would have been very high because the ozone layer that protects us today was not present. Dr. Robert Richmond, a NASA research biologist, stated that *D. radiodurans* beginnings are thought to be from the early Earth.

The ocean floor is one of the most extreme and violently active places on Earth. Huge tectonic plates rip and tear at each other, sometimes colliding or pulling apart, and sometimes grinding against each other as one slides along the edge of the other. The ocean floor was long thought to be a biological desert, completely barren of life. Submersibles capable of withstanding the enormous pressures of the deep ocean, however, have allowed biologists to visit the ocean floor, and to their surprise and delight, they have found a thriving community of animals and microbes.

Hydrothermal vents, tubes rising above the ocean floor, abound in some areas of the ocean. Pouring out heated water that is black with minerals, they have been called black smokers. Despite intense heat, pressure, and complete darkness, these vents often abound with crabs, tube worms, microbes, and other organisms which live in a world where no photosynthesis occurs. The minerals spewing from the vents serve as the basis of the food chain. On June 23, 1993, military undersea microphones picked up a seismic event of considerable magnitude along the mid-ocean ridge off the Oregon coast. A joint American-Canadian team of scientists sent to investigate saw clouds of small puffy white clumps of material being spewed out of a newly formed fissure. Looking much like a heavy snowfall, the white clumps turned out to be bacteria which apparently had been living deep within the ocean floor.

Many biologists now believe that life first arose near volcanic vents on the primordial ocean floor. Under this hypothesis, scientists suggest that the minerals and other chemicals at these vents would have simmered for untold millennia. Chemical reactions would certainly have taken place in these pressure cookers, creating new combinations that ultimately could have resulted in the first forms of life. If that is the case, why is this not happening today? The answer is that it probably is happening today, but microbes quickly gobble up any organic compounds being produced.

Scientists are able to synthesize simple organic molecules in the laboratory under the conditions thought to have existed on the early Earth. They can then polymerize these into more complex molecules like proteins, sugars, and fats, but that is a long way from producing life. The next step in the process would be to produce aggregations of biologically active organic molecules within some kind of a membrane. By controlling the movement of materials into and out of the proto-organism the membrane makes it possible to maintain a more or less stable internal environment that is different from what is outside. The membrane also prevents the loss of important compounds, some of which can be stored for future use. This ability is critical to cellular life, and would have developed early on or the complex molecules being produced spontaneously would have broken down as quickly as they formed. The cell membrane is selectively permeable and can maintain an internal environment that is radically different from the one outside the cell.

None of the steps described above need stretch the imagination very far to be accepted as a logical sequence for the spontaneous development of life on the early Earth. But to produce a living organism, some method of self replication is necessary. In other words, DNA, or perhaps RNA, must have first evolved, and along with these giant molecules, the enzyme systems that make self replication possible must also have been present.

This then is life, Here is what has come to the surface after so many Throes and convulsions.
- Walt Whitman

The DNA molecule (Deoxyribose Nucleic Acid) is the key to all life. Ancient beyond reckoning, it is a very large and extraordinarily complex molecule. Within DNA lies the genetic blueprint for all life forms. DNA has two capabilities which are absolutely critical to the formation and successful continuation of life: the ability to self-replicate; and, the ability to carry genetic information accurately from one generation to the next. Today's life forms represent billions of generations of faithful replication of the DNA molecule.

DNA, or a similar molecule, must have been present at the beginning of life, for by definition, life cannot exist without it. The development of living cells containing DNA was a quantum leap, and yet, if the fossil record is being interpreted correctly, it occurred more than 3.5 billion years ago. The structure and function of DNA are well enough understood today that a technology exists to create new life forms in the laboratory, and recently the human genome (the complete blueprint of all human genes) has been worked out. The first accurate description of DNA structure was unraveled in the early 1950's by Rosalind Franklin, Francis Crick, James Watson, and Maurice Wilkins. Based in large part upon Franklin's work Watson and Crick published the first accurate description of the structure of the DNA molecule in 1953 in the Journal Nature. They described a molecule which consists of two intertwined ribbons of sugar-phosphate molecules, looking something like a graceful spiral staircase. The connecting "steps" of the two ribbons are four nitrogen-rich compounds called bases: adenine, thymine guanine, and cytosine. Each DNA molecule contains hundreds of thousands of nucleotides in a long spiral chain. Watson and Crick postulated that the ability of the molecule to replicate itself is based upon the specific pairing of the nitrogen bases. Because an adenine nucleotide will only pair with a thymine nucleotide, and a guanine with a cytosine, when the molecule opens up, that is, splits down the middle, each half of the molecule can rebuild the missing half with the appropriate nucleotide units and thus two completely identical molecules will result, each of which contains half the original molecule.

Accuracy in replication is made possible by the specific pairing of the nucleotides with each other: adenine with thymine; guanine with cytosine. This is a critical aspect of the replication process, because it is the sequence of nitrogen bases along the molecule that determine inheritance. Changes in the sequence (mutations) will alter the genetic information being passed on to the next generation.

Watson and Crick's paper stimulated a tidal wave of research into the workings of DNA and its associated molecules that is unmatched in the history of biology. Within a few years, thousands of papers a month were being published dealing with DNA research. It wasn't long until the process of genetic information transfer was understood. The information for eye color, body height, and bone cell structure, etc. is coded into discrete sections of the DNA molecule, which for simplicity, we shall call genes. Each gene is a series of nucleotides within the DNA molecule, and this series of nucleotides spells out the "blueprint" for the building of a protein molecule.

Proteins are long chains of amino acids hooked together like the cars in a long railroad train. There are 20 different naturally occurring amino acids. Living cells are totally dependent upon proteins for their structure and function. Much of the structural materials which make up cells are proteins. Other proteins act as enzymes, and still others as hormones. DNA controls the synthesis of all proteins, and these in turn control the structure and functioning of the cell and organism.

The potential number of different proteins is essentially limitless. For example, if the average protein contains 200 amino acids, the number of different combinations is 20 raised to the 200th power. Consider that 20 to the 7th power is over one billion different combinations, and that 20 to the 8th doubles the number of combinations, and that the number will double with each digit added until 20 to the 200th is reached and you get some inkling that the number of different proteins that can be formed is greater than the number of atoms in the visible universe. It is the unlimited diversity inherent in the formation of protein molecules that is the raw material of evolution.

Most metabolic processes, such as the biosynthesis of molecules (hemoglobin, for example) occur in pathways where each step is facilitated by a specific enzyme. Thousands of chemical reactions occur in living cells, each one of which is regulated by a protein enzyme of specific structure. The action of an enzyme on its substrate is likened to a key fitting only one specific lock. The DNA molecule controls all of the activities of the cell by controlling the production of proteins, especially enzymes. They in turn regulate directly all physiological or structural changes within cells, which in turn regulate changes in the life cycle such as development of an individual from a fertilized egg; growth; life cycle changes such as metamorphosis, sexual development, annual rhythms; and behavior.

It is Written

I am a poem written in an alphabet of four letters.
Adenine thymine guanine and cytosine.
I am a skein of letters twisted and helixed.
My alpha and omega writ deep in my cells.
God wrote my blue eyes.
My sullen moods.
My love of tales.
My beginning
And my end.
Adenine thymine guanine and cytosine.
The letters of the Alphabet,
That wrote the epic of the whale,
The thrush's sonnet,
The viral haiku.
Man and beast and plant,
Poems in the same language.
Truly, God is the best of writers.

Eve Merrick-Williams

DNA Replication

The most crucial and difficult event to explain in the origin of life is the formation of self-replicating molecules like DNA and RNA. It is difficult to see how their development could have proceeded in a step wise process from simpler molecules. The DNA molecule controls its own replication; in other words, it can make an exact copy of itself. When a single bacterial cell or a cell in a plant or animal is about to divide to produce two cells, they must first replicate their DNA so that each of the resulting cells has a complete copy of the genetic blueprint. As is the case with most chemical processes, DNA replication is under strict supervision by many enzymes, all of which are coded for in the DNA molecule itself. M.C. Escher's drawing of two hands sketching themselves is an allegory for the ability of DNA to replicate based upon instructions contained with the molecule itself.

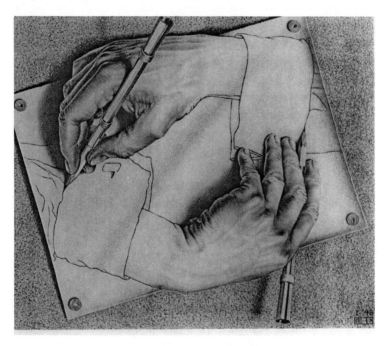

M.C. Escher's "Drawing Hands"©2008 The M.C. Escher Company-Holland. All rights reserved. www.mescher.com

The degree of precision in the replication of DNA is extraordinary. Nucleotide errors occur less than once in a million or so nucleotides. The fossil record indicates that some creatures, particularly marine animals, have remained unchanged in appearance for scores of millions of years. Horseshoe crabs, for example, arose about 200 million years ago and have remained relatively unchanged in appearance during that time. What other information transfer system could function with such accuracy during two hundred million replications? Imagine typing Shakespeare's Hamlet two hundred million times in succession, each time using the new copy to type from. A few errors would creep into the manuscript each time it was reproduced until it would be unrecognizable.

Despite the extreme accuracy of DNA replication errors do occur. If these were allowed to propagate the integrity of the genetic code would be compromised over time. Natural selection prevents this from occurring by weeding out errors that are maladaptive. On the other hand, some "errors" in DNA replication result in mutant organisms which are better adapted than the older forms and which eventually replace them. Although only a miniscule percent of all mutations are adaptive, mutations are the raw material from which natural selection draws in maintaining the fitness of the species. Without mutations the introduction of new materials, structures, and physiology would cease and evolution would be at a standstill. The eventual result would be the extinction of all life.

The most profound aspect of DNA is that its genetic code is universal. Bacteria, jellyfish, slime molds, maple trees, people - all living things (with only very rare exceptions) speak the same genetic language. This makes it possible to take genes from any one organism and splice them into the genome of any other. Not only will the transplanted genes function, but they will be reproduced with the host genes. Consequently, a transfer of genes need take place only once to be successfully propagated into the future. Firefly genes have been spliced into tobacco plants: the plants glow in the dark. Genes for human insulin have been transferred to bacteria, which now produce human insulin. Natural insecticide genes have been transferred from bacteria to a number of agricultural plants, producing the genetically modified crops (GMO'S)' that are so controversial in some parts of the world today. The universality of the genetic code is one reason most biologists believe that life originated only once, or, that if it had multiple origins, only one line survived.

The basic cellular components and metabolism of all living organisms are remarkably similar, and they function today pretty much as they did hundreds of millions of years ago. It is DNA *and* natural selection which make such stability possible. As mentioned earlier, horseshoe crabs and many other organisms appear not to have changed in hundreds of millions of years. During those same millions of years the Earth has undergone the most radical changes - mountains have risen and decayed, glaciers have grown and died, oceans have covered continents and receded, and continents have collided and separated. No physical feature of the planet can resist the decaying forces of nature as well as life; none can regulate and sustain itself. That life should exist at all is a matter of great fascination. That it should be so enduring passes the outer limits of imagination. Individual life forms are transitory and fragile, but life itself endures every catastrophe. It is undoubtedly the most enduring thing on earth.

And God said, Let the earth bring forth grass, the herb yielding seed, and the fruit tree yielding fruit after his kind, whose seed is in itself, upon the earth: and it was so.
Genesis 1: 11

There is no greater mystery to science than the origin of life. For people of faith who believe that God created the life forms of the Earth directly, this is not a problem. The quotation above is the first reference to life being created by the hand of God. In verse 20 God says, *Let the waters bring forth abundantly the moving creature that hath life, and fowl that may fly above the earth in the open firmament of heaven.* In verse 21 God creates great whales. It is not until verse 26 that God begins to think about creating man, which He does in verse 27.

What has been learned from these scriptures? Very little. This approach to the origin of life is an intellectual dead end. Having said that God created the Earth and all the living creatures on it makes any further investigation pointless. Science has done no better in solving the mystery of how life originated. To date scientists have not been able to present a convincing account of the development of life on Earth, and consequently do not have a credible argument to rebut the belief in special creation. Loren Eiseley, in his book *The Immense Journey*, pointed to this fact:

"With the failure....of many efforts (to unlock the mystery of the origin of life) science was left in the somewhat embarrassing position of having to postulate theories of living origins which it could not demonstrate. After having chided the theologian for his reliance on myth and miracle, science found itself in the unenviable position of having to create a mythology of its own: namely, the assumption that what, after long effort, could not be proved to take place today had, in truth, taken place in the primeval past."

Science does not understand how life originated any better than it understood the causes of disease before Pasteur developed his germ theory, but it does have a powerful method for investigating natural phenomena which it can bring to bear on the problem. It is a mystery three and a half billion years old. We will never truly identify our place in the universe until the riddle of the origin of life on Earth is determined conclusively by either science or religion. While a few people on the outer fringes are satisfied with beliefs that cannot be verified, the overwhelming majority of people are not.

It is my firm belief that when we are able to accept what and who we truly are, the natural products of an extraordinarily dynamic and creative planet, one of millions of species whose existence is interdependent with all the others, we will begin to conquer the fear that has been a hallmark of our species. We will then come out of the darkness of our primitive night into the brilliance of a new age.

Humanity Unveiled

If we are to understand who and what we are we must first understand the long road traveled in our evolutionary history. There is much misunderstanding today about the scientific theories with which biologists and anthropologists interpret our natural history. Some of the misunderstanding results from a deliberate effort to confirm some special interest or religious belief. Scientists make no claim to absolute truth when it comes to their own theories; quite the contrary, they see them as works in progress which must change as new interpretations become available. What follows in the next five chapters is a review of our still incomplete understanding about the biological history and nature of humankind.

CHAPTER 9

Hands

Beauty comes in knowing what you are and whence you came and why you be, earth child.
- Daniel Kozlovsky

A small female stumbled down through the dense vegetation covering the hills surrounding a lake. She had no name but would later be named Lucy. Lucy was in trouble, very serious trouble, for her unborn child was trying to enter the world but the delivery was not going well. Agonizing pain swept over her in waves. She was alone and very frightened as she made her way toward the lake, hoping to find there some of her family and friends. There was no one, and she sank down onto the shore, screaming out her agony in short gasps. She lay there through the heat of the day and into the night, her small body convulsing with the effort to bear her child. At dawn, as the first faint rays of morning light illuminated her face, the child was stillborn.

Lucy picked the little one up and gently wiped away the blood from its tiny face. Looking into the lifeless brown eyes for a moment and thinking of what might have been, she lay back with the child beside her and let the morning sun warm her face. She was dazed and totally exhausted from the effort and loss of blood. Dimly, she heard something coming through the bushes and turned to see a pack of hyenas approaching at an easy trot. Despite her weakness she leaped to her feet and looked for a way to escape. There was none, and she ran into the warm water until it was up to her neck.

The hyenas approached the body of the child and began to squabble over it. The female, leader of the pack, took possession of the infant and the others backed off at her warning snarls. Lucy watched in despair but could do nothing. Several of the other hyenas came to the water's edge and threatened Lucy, but did not enter the water. They lay down, watching her, panting with mouths agape, willing to wait for a meal. Involuntarily Lucy backed away from them and lost touch with the lake bottom. Frantically she tried to reach it with her toes, but she had gone too far out and didn't have the strength to swim back. For a few moments she was able to keep her head up, but in her weak and exhausted condition she quickly gave up the effort and slipped beneath the water. As the water closed over her she relaxed, and her pain and terror subsided.

Lucy's body decayed away in the weeks and months that followed, and the bones settled into the lake silt and were covered up. They would not see the light of day again for more than 3 million years.

Luck is a Lady Tonight

By his own admission Donald Johanson is a superstitious man. He has reason to be, for success in his profession is largely a matter of luck. Johanson is an anthropologist who has spent his career searching for some of the rarest and most precious objects on earth – fossils of human ancestors. One hot day in November of 1974 Johanson set out from his camp site in the Afar Depression of Ethiopia to search the dry, bare hillsides for fossils. The previous year at the same site he discovered a knee joint which was unmistakably human, but which was found in a layer of soil at least 300 million years old. The find electrified the scientific community and made it possible for him to raise the money to return a year later with a team and search for fossils of the creature whose knee he had found.

As he left camp that morning Johanson had a premonition that something special was about to happen. For several hours he walked the bare rocky hills intently staring at the ground. He saw many mammalian fossils, for the area is rich in them, but none were what he had come to find. After searching for several hours he decided to return to camp for lunch when suddenly he spotted a piece of bone sticking out of the rocky ground. Picking it up, he discovered that it was part of a hominid arm bone. Johanson and his team later recovered 40% of a complete skeleton, an almost unheard of prize in anthropological circles.

The whole team worked late that night cleaning and examining the fossil bones. As they worked, the realization that they had discovered one of the great finds of all time led to a celebration with drink, dancing, and song. The Beatles' song *Lucy in the Sky with Diamonds* was played over and over, and thus did the skeleton gain a name and something of a personality.

Lucy was later given the scientific name *Australopithecus afarensis*, which means African Ape from Afar. She was a small ape, about 3 ½ feet high, weighing about 60 pounds and with a brain not much larger than a chimpanzee's. She was approximately 25-30 years of age at her death. What set her apart from all other apes was her upright posture. This uniquely human characteristic in a small-brained animal that lived more than 3 million years ago makes it clear that our lineage split away from the other great apes through walking upright, not through an increase in brain size. Later fossil finds have confirmed this conclusion.

Lucy and her kind evolved from apes which moved through the forest by swinging under limbs rather than running on top of them as monkeys do. To hang from a limb requires a hand which can close around the limb and give a secure grip. For this purpose the thumb curls in an opposite direction to the other fingers. When walking on the ground chimps, orangutans, gorillas and the other great apes walk on all fours, using the knuckles of their hands for support. They can stand partially upright, but the structure of their feet, legs, and pelvis prevent a complete upright stance, and being unable to lock their knees as humans do when we walk, they shuffle along with knees bent.

Scientists are sometimes no different from other people in allowing their preconceptions to mislead them. In the early part of the 20th century (between 1908-1912) a skull and jaw were found in England which was given the name Piltdown Man, and which was heralded by scientists as being a true ape-man, thus bridging the gap in the fossil record between humans and apes. The skull was the same as that of a modern human, while the jaw was that of an ape. Piltdown man was revealed to be a fake in 1953, much to the embarrassment of the anthropological community that had endorsed it for more than four decades. The hoax was perpetrated by a person or persons unknown who planted a modern human skull with the jaw of an orangutan. The scientists studying human evolution at the time were so certain that the transitional form from ape to pre-human would be a "brainy ape" that they immediately fell for the Piltdown hoax. There may also have been a certain amount of nationalistic pride involved. The few hominid fossils found up until the time of Piltdown had all been found in Africa and Asia, and finding a fossil in England, the home of Darwin, Lyell, and Huxley was enormously pleasing to English scientists.

The Piltdown fraud retarded early research on human evolution by leading scientists down a blind alley. Discovery of authentic anthropoid fossils found in South Africa in the 1920's were dismissed or ignored because unlike Piltdown they were small brained. An enormous amount of time and effort were devoted to the Piltdown fossil (more than 250 scientific papers were produced) which could have been directed toward more productive studies.

Walking Upright

The discovery of very early hominid fossils like Lucy destroyed the "brainy ape" theory completely. Our earliest ancestors were apes with brains no larger than that of modern apes like chimpanzees and bonobos. What made them different from all other apes was that they were bipedal. Upright walking was the initial impetus for the development of our species, and the big brain came much later in our evolution.

Walking upright made it possible for apes to live where apes could not live before.

About five million years ago the forests in Africa began to disappear due to climatic changes. They were gradually replaced with woodland and savannah – grasslands with scattered trees. Many animals, primates included, which had lived in the deep forests became extinct, but some, like Lucy's species, were able to adapt to the change and move out, at least for part of the time, into the more open areas. Apparently, walking upright was an advantage in the new environment, but there were also serious disadvantages. Most primates avoid predation by climbing trees or fighting with their large canine teeth. Both of these advantages are lost with an upright posture. Putting the canine teeth (fangs) on top of the body instead of in front of it takes away their usefulness in fighting, and in the course of evolution they began to shrink in size (Imagine trying to fight a dog with your teeth. To do so you would have to get on your knees!) The large root of the canine tooth, however, remains to this day to remind us of our ape ancestry.

Apes are flat-footed with the great toe separated from the other four toes so that the foot can be used like a hand in climbing and grasping objects. Lucy's foot was more like that of a human. Her big toe was in line with the others, which created an arch for the foot to absorb the shock of walking or running. Thus, the foot was no longer of use in climbing.

Another serious disadvantage of an upright posture is that it reduces the size of the birth canal. The canal through which a child is born is completely surrounded by the bones in the lower part of the pelvis. Narrowing this opening made birth more difficult. This was probably not a problem for Lucy since her brain was not much larger than a modern Chimpanzee's, but later in human evolution when the brain began to increase in size the problem became more acute. The larger brain meant a larger head which had to pass through the birth canal.

Children of God: Children of Earth

BIZARRO (new) © DAN PIRARO. KING FEATURES SYNDICATE

The difficulty of childbirth was eventually solved by shortening the gestation period and giving birth to a very underdeveloped child. Compared to other apes, human infants are born very prematurely, which makes the child totally helpless for a prolonged period following birth. Given these serious problems with standing upright, there must have been overriding advantages or natural selection would have eliminated individuals which adopted the upright posture. Several advantages have been proposed. Walking is very efficient compared to other forms of locomotion; it requires less energy, and great distances can be covered tirelessly in a day. Living in the forest and feeding on leaves and fruit which are readily available, apes do not have to go long distances to find food, it is all around them. Things are very different on the savannah, where food is sparse and widely scattered. Individuals who could cover large distances more efficiently would be favored by natural selection.

It has also been suggested that standing upright is advantageous in grasslands because of the ability to see over the grass. Finding food and avoiding predators is easier with the head on top of the body rather than in front of it, and with the hands free it is possible to carry food back to a den or home site. Free hands also make it possible to carry a stick or bone as a weapon to ward off competitors or predators. Chimps occasionally pick up sticks to use as a weapon but they must throw them away when they drop down on all fours to run. Consequently they primarily use sticks or other objects without modifying them.

Moving from the cool forest to the extremely hot savannah would have created problems in keeping the body cool. A temperature rise in the brain of as little as 5 or 6 degrees is enough to cause heatstroke and death. Dr. Peter Wheeler, a physiologist at Liverpool John Moores University in England has calculated that an animal Lucy's size would be exposed to 60% less solar radiation walking upright than on all fours.

All of these factors – the ability to cover ground efficiently, free hands to carry food and other objects, seeing over the grass, and keeping the body cool were apparently sufficiently advantageous in the new habitat and climatic conditions that natural selection favored them despite the problems involved in upright walking.

Chance Favors the Prepared Mind
- Louis Pasteur

One of the most incredible and exciting finds in the search for human ancestors was reported by Mary Leakey at a place called Laetoli in Tanzania. Several hundred thousands of years before Lucy lived, farther south in Africa two others of her kind walked side by side across wet volcanic ash. As they walked birds flew up before them and a hare ran away. The birds, hare, and two hominids all left their footprints in the moist ash, and as ash continued to fall from the sky it covered the footprints. Later deposits and time converted the ash to a cement-like hardness which protected the footprints. They were discovered by two men in Mary Leakey's team of anthropologists as they playfully threw elephant dung at each other. As one of the men dived out of the way he fell upon the footprints and recognized their importance.

The foot prints are at least 3.6 million years old and are hardly distinguishable from human footprints. Looking at the two sets of prints which lead away in the stone for 80 feet, one can picture two people walking hand in hand, but they were not modern humans, they were Lucy's people – small, upright apes who may or may not have been the progenitors of our species. Fossil bones of Lucy's kind which are more than four million years old have also been found at Laetoli by Leakey's team.

The Laetoli footprints show very clearly the modern form of walking: the heel strikes first, followed by the rolling of the foot forward over the arch, and finally the push with the big toe. An ape's footprint, on the other hand, shows a thumb widely separated from the other toes, and a flat print caused by planting the entire foot rather than striking with the heel.

Everything that we are as a species followed the development of an upright posture. Ask any 100 people on the street what it is that sets us apart from all other creatures and most will reply that it is our big brain. Not so. What sets us apart is our upright posture and the freeing of the hands from walking and running. Without hands the big brain would never have evolved, nor would language, science, the arts, religion, agriculture, or anything but the most rudimentary technology. We have named ourselves *Homo sapiens*, meaning "Wise Man" in Latin. We should have named ourselves *Homo habilis,* "Handy Man."

In the millions of years that separate Lucy and her kind from modern man standing upright lead to changes in many parts of the skeletal system. The backbone changed from straight or curved to S-shaped to absorb the shock of walking and running and to place the center of gravity of the body directly over the feet. Most of the changes, though, were in the skull. Placing the head on top of the spinal column instead of hanging from the end of it as it does in four-legged animals required moving the foramen magnum, the opening through which the spinal cord attaches to the brain, from behind the skull to under it.

When the skull hangs from the end of the spinal column, as it does in animals that walk on all fours, the heavy weight of the head requires powerful muscles to hold it up and to use it in the grasping and shaking that goes on during a fight. These muscles attach to the back of the skull, which is flat. At their other end the muscles attach to long projections from the bones of the vertebrae. Hold a book at arm's length for any period of time and you will appreciate what it is like to carry a heavy skull hanging out in front of the body. Positioning the skull on top of the spine made it possible to do away with the flat plate at the back of the skull, and reduce the size of the muscles and their attachment points on the spine. It was then possible for the skull to become thinner and lighter and to support a larger brain.

The lower jaw of an ape is massive compared to ours. Huge muscles attach to the jaw, run up under the cheek bone, and attach to the sides and top of the skull. To support the weight of the jaw and muscles, and to keep the skull from being deformed by the pull of the powerful muscles, a ridge of bone stands upright across the top of the ape skull (like the top of an old Spanish Conquistador helmet) and thick eyebrow ridges protrude over the eye sockets. This was how Lucy's skull was structured, but in the course of human evolution the loss of large canine teeth for fighting lead to a much reduced lower jaw and the muscles which support it, and a pulling back of the snout into a flat face. The heavy eyebrow ridges and crest on top of the skull also disappeared since they were no longer of any use. These changes greatly reduced the weight of the skull and eventually made possible increases in the size of the brain. These changes took millions of years to complete but the result was a much lighter skull with the potential to support a larger brain.

If you have difficulty accepting that changes in the skeletal system of the kind I have described could occur in the human lineage, consider the changes which have occurred in the evolution of wolves into dogs as varied as Pugs, Dobermans, and Sharpeis. These changes took only 12,000 years or so under the intense selection of their human domesticators. There is at least as much genetic diversity in the human gene pool for evolution to draw upon as in the gene pool of the wolf. In his book "Days of a Thinker" Loren Eiseley describes striking up a conversation with a huge powerful man he met on a train from New York to Philadelphia. The man had a double thumb and his fingernails were like the claws of an animal.

The man's size, extra thumbs, and claw-like fingernails are all outside the mainstream of human characteristics, and all are most likely due to genetic differences. What we consider to be normal regarding the human body is a model that exists only in our minds. Consider the following people from the Barnum and Bailey sideshow:

Lucia Zarate: born in Mexico in 1864. Weighed 8 ounces at birth and was 7 inches long. As an adult she was only 20 inches tall and weighed 5 pounds.
Stephan Bibrowsky: Born in 1890 in Poland. Dense six inch long hair completely covered his face and body.
Al Tomainia; Born in New Jersey in 1912. He was 8 feet 4.5 inches tall.
Simon Metz: Microcephalus (very small head) – mental age of 3.

The list of exceptional traits in humans would be extraordinarily long. It would include contortionists who can bend their bodies into pretzels, children who can solve complicated math problems or play Classical music with no formal training, people with club feet, cyclopia (only one eye), or a host of other exceptional traits. The New England Journal of Medicine reported in 2004 on a 4 year old child in Germany who had muscles twice the size of other children his age and half the body fat. Scientists were able to identify the DNA mutation responsible for the boy's condition.

Tanya Streeter, a multiple world record holder in freediving was called "The World's Most Perfect Athlete" by Sports Illustrated in 2002. Her lung volume is almost twice the normal for a woman of her size and build. In 2002 she made a dive to 525 feet and returned to the surface on a single breath of air.

Does the genetic diversity exist in the human gene pool to produce a race of nano-people: people only 5 inches tall? Clearly, the answer is yes. How about a race 8 feet tall? Yes again; the only thing needed is for selection to take place for small size or large size over many generations. In a sense we are all mutants, for each of us is unique genetically, and each of us on average is carrying 3 mutations that were not carried by our parents. These accumulate over generations until we each are carrying on average 300 or so mutations. Mutations are normally harmful because they are alterations of successful genes and are therefore maladaptive to current environmental conditions, but when conditions change today's harmful mutations can become tomorrow's savior of the species.

A case in point is a mutant gene that causes red blood cells to take the shape of a sickle. As such they have difficulty flowing smoothly through tiny capillaries and can shut off the supply of blood to major organs, causing injury or death. The cells are also fragile and break apart easily, which reduces the bloods ability to carry oxygen and results in anemia which can cause heart failure. People carrying both copies of the mutant gene will die; those carrying only one can suffer serious injury and can also die from its effects. The gene is therefore kept in low frequency by natural selection within most human populations.

In some parts of Africa, the lowlands around the Mediterranean Sea, and parts of the Indian subcontinent the incidence of sickle cell anemia is high. The reason for this apparent contradiction is that people with the mutant gene are more resistant to malarial infection than normal people. In places where malaria is a greater threat to survival than sickle cell anemia, the gene is selected for and increases in frequency. Among U.S. blacks and peoples whose ancestors came from lowland regions of Italy, Cyprus, Greece, Sicily, and the Middle East, 1 in 12 individuals is carrying this harmful mutant gene.

Any change in the physical or biological environment of a species can alter the selection pressure on genes and cause a former harmful gene to become useful and more frequent.

> Deformity is daring.
> It is its essence to o'ertake mankind
> By heart and soul, and make itself the equal –
> Aye, the superior of the rest. There is
> A spur in its halt movements, to become
> All that the others cannot, in such things
> As still are free to both, to compensate
> For stepdame nature's avarice.
> *Lord Byron*

Lord Byron was not speaking of the role of mutations in the evolution of species but of his own deformity and the way physically challenged people can compensate and achieve success; however, his lines could just as well be addressing the importance and role of mutation in natural selection.

Lucy and her kind faced a monumental challenge in moving out of the forest and into the savannah. Had they failed to make the transition, as many other species did, we would not exist today. Lucy's kind successfully battled the forces of extinction for over four million years, a record which our species seems unlikely to match.

Lucy's people could not have wandered out onto the savannah unless they were adapted to meet the challenges of that new habitat. It is generally true that plants and animals cannot change habitats at all unless they are minimally preadapted to the new habitat. Fish could not move out of the water and onto the land, for example, until they could breath air, have structures for moving around, be able to find food that was edible and digestible, survive the more extreme and rapidly changing temperatures, and have sense organs which could function in air rather than water.

We know from the fossil record that fish did make the transition to land dwelling amphibians, but why would a fish preadapt itself for life on the land. Does this imply that evolution is forward looking, that fish were planning for the invasion of the land or that it was a preordained move? Based upon evolutionary theory we have to reject such thinking. Evolution is not predictable or preordained. Evolution prepares living things for life in the present, not the future. Fish living in small pools or ponds of water which dried up periodically may have developed the ability to breathe through their skin, as amphibians now do, in order to move from a pool that was drying up to one that still contained water, To do so they would also have to be able to slither along in the wet mud, and their senses of sight and smell would have to guide them to another pool.

With time, the crude legs would develop into more proficient ones, as would the senses and the ability to breathe through the skin. Eventually, these poorly adapted creatures would be able to leave the water for longer and longer periods of time. Natural selection would then take over and shape them into more complete land creatures. It is important to note that the pre-adaptations which allowed the change in habitat actually evolved so that the fish could move from one water hole to another: in other words, the pre-adaptations for land life were actually developed to enable the fish to stay in water.

This may or may not have been the sequence of events which took place as some fish invaded the land and became amphibians, but whatever the sequence was, the first land vertebrates were able to survive only because they had the land to themselves. Being poorly adapted at first (amphibians are still poorly adapted to land life in that they have never evolved an egg that can develop and hatch on land. They are required to return to their watery beginnings to reproduce.), there must have been intense selection pressure to improve.

Lucy's species must have developed bipedalism in response to challenges within the forest and perhaps at the forest edges. This would have allowed them to make forays into the savannah, and those individuals who were able to feed themselves and avoid being killed and eaten out there would have left offspring, some of whom would have had their parents' survival traits. One of the most important preadaptations for life on the savannah was a hand with an opposable thumb. The hand evolved for grasping tree branches while swinging through the trees, but it worked extremely well at carrying food, weapons, and tools on the savannah.

There is much disagreement as to the line of descent from the earliest hominids to modern man. Lucy may have been our direct ancestor or merely one of the side roads along the way. The number of hominid species represented in the fossil record continues to grow as anthropologists make new finds, and it is difficult to sort through them and find evidence for a direct line of descent. There were obviously many forks in the road once bipedalism had been established. The modern era is unique in having only one hominid species extant on the planet at the same time. During most of our evolutionary history multiple species of hominids coexisted. This ended with the disappearance of the Neaderthals some 30,000 years ago. We are the lone surviving relatives of the first ape to walk upright.

The hand also changed in important ways to give it the ability to do more skilled work than is possible with an ape's hand. The thumb grew longer, rotated freely, and became more opposable to the other finger tips. As these changes in the skeleton were occurring certain areas of the brain began to increase. Giving the thumb and index finger greater importance in survival required better communication between the brain and these two fingers. This meant increasing the size of those areas in the brain that controls them.

Last of all came the increase in those areas of the brain that we associate with rational thought and intelligence. Explaining these changes is much more difficult than explaining the selection pressures that forced development of the modern skeleton from that of a humanoid like Lucy. Natural selection has no tolerance for waste or inefficiency, so doing away with heavy massive skulls, teeth, and muscles makes perfect sense because they are costly to build and maintain, but behaviors can also be expensive to maintain and the benefits must outweigh the costs or selection will eliminate them or replace them with something more cost effective. Music, art, and religion are all costly in terms of the energy we devote to them, and they appear to be as old as our species. Biologists find it difficult to explain how these behaviors could have evolved.

I wish I could pierce the mists of time, could stand by Lucy's lake and say something to her in her last hour, but what would I say? Her life, like mine, was filled with pain and joy, defeat and occasional success. She wouldn't understand my words, of course, but her genes are in me, and but for her willingness to fight against all the forces that sought to destroy her I would not be here today. Perhaps the kinship that we share would allow me to convey my thanks to her in other ways, a knowing look in the eyes, a shrug of the shoulder, or, if she would allow it, a pat on the back. Her life has meaning for me and for all of the people alive today. But for her, we would not exist. She and her kind took a chance few of us would be willing to take. Leaving the security of her home in the forest, she ventured out onto the hot, dry grassy plains where food was scarce and the world's most fearsome predators prowled. But for her skill and will to live life could have ended there for us all.

CHAPTER 10

Handy Man

The man crept silently into the dark cave. In his hand was a small stone which had been hollowed out to hold animal fat. This was his lamp. The light from the lamp flickered, its yellow light casting a kaleidoscope of moving shadows around him. He placed his feet carefully, and warily looked about as he moved deeper and deeper into the cave. The darkness frightened him, as it did all members of his kind, and his body was taut with the tension.

Suddenly a shaggy beast appeared before him on the wall. He held the lamp higher, revealing animal shapes all around the stone walls. After carefully placing the lamp on a tall rock, he examined each of the painted animals with the reverence of a Jew visiting the Wailing Wall or a Christian following Christ's path to the cross. This was his holy place, a place where he communed with the spirit world. His was a world of animal spirits, and by depicting them on the wall of the cave, he was reaching out to them, for without them he and his kind could not continue to exist on the cold, glacial plains of Europe. This was ice age man communing with the gods of his world. This was his prayer.

Taking out a small leather pouch which contained his pigments, he mixed them in turn with cave water or saliva, and began work on the painting of a small horse. Hours went by, during which he became completely engrossed in his work so that the fear of the darkness was lost to his mind, but every now and then some small noise in the cave would bring him around and he would stiffen and glance wide-eyed into the darkness.

When he had finished the drawing, he backed away and held the lamp high to examine it. After making a few more touches on it he grunted his approval, then ground up more charcoal with his small stone mortar and pestle, and put a small handful of it into his mouth. He chewed on the charcoal making crunching sounds as he broke up the small pieces with his teeth. When he was satisfied that he had the proper consistency, he held his hand to the cold stone wall and spread his fingers. Leaning close, he began a rapid spitting motion – t-t-t-t-t-t-t-t- spewing small bits of wet charcoal around his hand. He continued until his fingers and the edges of his hand were black with it, and then slowly backed away. He stared at the handprint on the wall for long seconds: it was his signature and his immortality. He was reaching out to all who would follow him. He was saying I am a man, but he was also saying I am a man among men. I am me.

As late as the middle of the last century it was generally assumed by scientists and lay people alike that our big brain was the key to our survival as a species. Not so. Our humanity came with our hands. Like our ancient ancestor, our hands are the windows to our soul. More than any other part of our body, more even than our face, our hands reflect who and what we are as individuals and as a species. We are extraordinarily proud of our big brain, but the brain is an extension of the hands. In our evolution as a species our hands came millions of years before the big brain, and hands are largely responsible for the increase in brain size and complexity. When our ancient ancestors left their hand prints on the walls of caves in France and Spain they were expressing the importance and individuality of their hands to themselves and to their society. Only very rarely did they depict people, and when they did the paintings were mere stick figures with none of the detail or accuracy of the animal paintings. The only accurate depictions of the human body commonly found are the hand prints. It was their signature as individuals; it was a representation of their pride in what they had accomplished and were capable of accomplishing with their hands. Modern artists, including photographers, focus primarily on the human face, but hands are as expressive and individualistic as faces, and to ancient people whose everyday existence was so directly related to what they could do for themselves with their hands, the face apparently meant very little. Religion, language, art, music, rational thought, technology - perhaps even sentience – indeed everything that we are as a species - are the direct or indirect result of our hands.

Every species has a unique survival strategy. If it works well they prosper in the ongoing competition with other species for the resources necessary to survive. If it fails, they decline and either improve the strategy or become extinct. The rules of survival are inexorable – only the best adapted survive. Looked at from a biological standpoint, the key element in our survival strategy is our hands and the changes in brain anatomy which resulted from the increasing importance of hand use. Our hominid ancestors survived for millions of years without our large brain; they would not have done so with clumsy hands.

Standing erect freed the hands from the knuckle walking of most apes. Once free, evolution could have made hands less important, like the tiny forelimbs of dinosaurs, but instead they took on an importance of such great value that it resulted in quantum changes in body structure, behavior, and culture. Evolution is opportunistic and unpredictable. No one really understands why it turns off one road rather than another, but once it has done so the commitment is made and the road will lead either to success or extinction. Along the way many side roads present the opportunity for further change in direction, some of which lead nowhere while others may be hugely successful for millions of years.

So it has been in the course of human evolution. Since the first ape stood upright there have been dozens of erect anthropoids, all of whom had hands with which to carry food, manipulate objects, make tools, and communicate with each other. We are the only survivors of that group. Thirty thousand or so years ago our closest relative, Neanderthal Man, disappeared, and for the first time in five million years only one species of mankind existed on the Earth. Anthropologists are trying to work out the direct lineage of our species from the fossil record, and new finds are being made regularly which add new evidence and alter older perceptions of our course as a species. What *is* clear are the changes which led to our development from a bipedal ape like Lucy. The course of these changes is represented in the anthropoid species described below.

Australopithecus afarensis. Represented by the Lucy skeleton and other fossil finds in east Africa. 3.9 - 3.0 mya (million years ago). Brain size 375-550 cc. The face was ape-like with a sloping forehead, bony ridge over the eyes, a flat nose and without a chin. The canine teeth are considerably smaller than in modern apes but larger than humans, while the shape of the jaw is intermediate between modern apes and humans. The pelvis, leg bones and feet resemble those of humans and make it clear that they were bipedal. Found only in Africa.

Homo habilis. Called "handy man" because crude stone tools have been found with the skeletal remains. 2.4-1.5 mya. Brain size 500-800 cc. About 5 feet tall and 100 lbs. The face is still very ape-like with a projecting jaw, large back teeth, and a flat nose. The bulge of Broca's area in the brain, which is necessary for speech, may indicate the capability of rudimentary speech. Found only in Africa.

Homo erectus. 1.8 million – 330,000 years ago. Brain size 750-1225. Face has protruding jaws with large back teeth, thick brow ridges, and a low, sloping forehead. Found in Africa, Asia, and Europe. Probably used fire, and made stone tools that were more advanced than *habilis*. *H. erectus* was much more robust and stronger than modern man and may have been a better walker. The face was shortened with a prominent nose. His height was about the same as a modern human. From the neck down *erectus* could have passed for a very robust human. His species ranged from Africa to Indonesia, China, Eurasia, and Western Europe. In all probability, both *Homo sapiens* and *Homo neanderthalensis* descended from *erectus* in Africa. The survival of this species for more than 1.5 million years is indicative of its highly successful anatomy and lifestyle.

Homo sapiens. 200,000 – present. Brain size 1350 cc. High forehead with very small eyebrow ridges. Prominent chin and lighter skeleton than *H. erectus*.

The above descriptions are not meant to indicate that this is the actual lineage of the human species; there is disagreement among anthropologists about the exact line of descent. What they do indicate are the gradual changes which took place over millions of years in hominid evolution prior to the emergence of modern man.

Evolution of the Human Hand

Standing erect was the direct result of having hands that could hold and manipulate objects. Without hands that could carry and grasp objects, standing erect would have meant quick extinction for any ape species that experimented with it.

Apes like chimpanzees are too large to walk or run on the tops of branches the way monkeys do. Swinging under the branches gives them much better stability while moving through the tree tops. The thumb of the great apes is very short compared to the human thumb and this makes it impossible for them to fully oppose the tips of the other fingers with their thumb. Apes are also incapable of rotating their thumb, which restricts opposability even further. Nonetheless, some chimps and other apes do make and use primitive tools.

Jane Goodall was the first person to observe chimpanzees using a tool. Chimps regularly eat termites, which are high in protein, low in fat, and are generally very nutritious. In order to capture termites, chimps use a stick which they poke into holes they have made in the termite mound. After waiting for a few minutes, they pull the stick out and eat the termites that are clinging to it. They hold the stick by closing their fingers around it the way you would hold a pencil if you had no thumb. A human performing the same task would hold the stick between the pads of the thumb and the other fingers, but this is impossible for chimps.

As John Napier has pointed out, our hands are capable of two different grips: a power grip, and a precision grip. The power grip is what we use to hold a hammer or a pistol. It involves curling the fingers around the handle in one direction and the thumb in the other. By opposing the fingers with the thumb great power can be exerted. This is the same grip that is used in "cracking" the tight lid on a jar – fingers opposing thumb. Once the lid is "cracked", however, we switch to a different grip in which the lid is unscrewed with the tips of the fingers and thumb. This is a called a precision grip because we can do much more careful and precise work with it. This is the grip one would use when sewing with a needle or doing any fine work. The hand of great apes is capable of a power grip, but is very poorly designed for a precision grip.

There is considerable variation in the hand structure of different apes. Gorillas have the longest thumb relative to the fingers; orangutans the shortest. Gorillas rarely leave the ground, while orangutans live in the trees. Most of the fruit and tender leaves which orangutans eat are at the periphery of a tree on the smaller branches. They swing from small branches by hooking their very long fingers over the limb without using their thumb. Chimpanzees, which spend time on the ground but also move into the trees to hunt for food, have a hand that is intermediate between that of a gorilla and an orangutan. Evolution has taken a hand that is very primitive in design (the five-digit hand and foot are at least 60 million years old) and adapted it in different ways for different primate life styles.

The alterations in the bones, muscles, and tendons which resulted in the human hand are important, but the hand does not operate alone. Shortening of the arms relative to the legs and improvements in the structure of the shoulder were also important in giving the human hand its extreme versatility. The hips are also important in the work of the hands, as anyone who has ever hit a golf ball or thrown a football knows.

Perhaps most important are the alterations that took place in the brain as a result of the new importance placed upon the hand with upright walking. Eye-hand coordination and cooperation between the two hands are functions of nervous control from the brain. Lucy's hands, and the hands of all the hominids which followed her, were being asked to do things not required of the ape hands from which they evolved. Some of these functions, like the making of tools, were critical to their survival, and greater control of the hands by the brain was required.

Once our hominid ancestors like Lucy were walking upright there was selection pressure to make the hand more effective in its new tasks. Lucy's hand was similar in structure to humans but the thumb was shorter and the fingers were slightly curved.

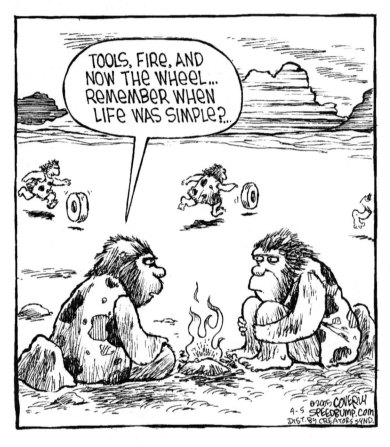

"Speed Bump" © Dave Coverly/Dist. By Creators Syndicate

> Man is a tool-making animal.
> - Benjamin Franklin

Chimpanzees and other apes use tools and occasionally make them, but only humans have staked their survival on a tool technology. Chimpanzees regularly throw sticks, bones, stones, and other objects, and they use leaves to clean themselves after defecating, but making a tool is very different than using natural objects for a short period and then throwing them away. A chimpanzee that breaks off a thin twig while on the way to a termite mound, then stripes off the leaves and trims the twig it to the proper length shows a degree of conceptualization not required in picking up a stone or stick to ward off a snake. It requires imagination, foresight, memory, and the cultural transmission of knowledge from adults to young. The tool maker usually makes the tool away from the place where it will be used, and he must consider what material to use, how the tool will respond to the forces it will be subjected to, and how large it needs to be. He then fashions it with the use it will be put to in mind.

Lucy's tools were very crude, not much more than natural stones, and she and her kind were no more adept at using tools than modern non-human primates. As the hand evolved through *Homo habilis* and *Homo erectus*, it became increasingly more capable of precision work and the tools produced consequently became more sophisticated and useful. This created the need for greater neural control of the hands and as tool use increased and became more critical to survival natural selection favored an increase in the size and complexity of the cerebral cortex to control the movement and use of the hands. Some of the increase in brain size was for better neural control of the thumb and fingers, while others were for the kind of memory, foresight, and imagination that allowed for new tools to be created.

From crude single edged stone tools, to finely made double edged tools, to putting the stone on the end of a stick to make a spear, to creating a spear thrower, to a bow that would throw a small spear even farther, to plows, guns, and refrigerators, our ability to feed and protect ourselves has steadily improved and our numbers grown as tool technology progressed.

Humans cannot repudiate technology; it is the key to our survival as a species. Neither can we lock it in place by foregoing innovation – technology has an irresistible force that propels innovation forward at exponential speed.

Our ancestor species survived and prospered for many millions of years without our enlarged brains. They did so because they stood upright and had our hands (or we theirs). With the increase in brain size in *Homo erectus* and a further increase in the evolution of our own species, the hand took on new functions and improved upon older ones.

With the eye the hand is the main source of contact with our physical environment. When we say "let me see it", what we really mean is "let me feel it while I examine it with my eyes." The importance of hand touching as a way of sensing our environment cannot be over emphasized. With the hands we can "see" in the dark or in small spaces which are hidden from our eyes. The probing fingers of a doctor; a mother's soothing touch with an ailing child; the feel of the controls in the hands of a pilot; the press of a ball as it comes into contact with the hands of a receiver; the intimate caress of a lover's hand; the strong (or limp) handshake of a stranger; these and tens of thousands of other uses for our hands we take for granted and do without thought, but the whole time information is passing from the hands to the brain and from the brain to the hands.

We also communicate with our hands; in fact, the earliest communication may have been through hand signals. A wave says goodbye or hello; a clenched fist is a warning, or two clenched fists in the air can signal triumph and elation; a shaking finger from a teacher or parent expresses anger or disapproval; two open hands held facing another person means whoa, back off, or take it easy. We use hand signals unconsciously but we react to them in a very conscious way. Hands express words, thoughts, ideas, and emotions, and they do so in a very unambiguous way.

Broca's area, that part of the brain in humans that has much control over the hands is also involved in speech, which makes it seem likely that the hands were important in the development of our ability to speak. It is not difficult to imagine our ancient anthropoid ancestors using hand signals to communicate while on a scavenging or hunting expedition on the plains of Africa. Not only would such signals help coordinate actions between individuals, but being silent they would not alert predators. It seems likely that in the course of evolution we "spoke" with our hands before vocal languages developed.

Humans are not the only animals with an enlarged brain. Whales, porpoises, elephants and apes are also very intelligent animals, but none of them has developed any kind of technology. Imagine if you can, a whale trying to build a crude raft in order to carry a youngster too sick to swim, or a chimpanzee making a net out of vines in order to capture monkeys. Both are impossible for different reasons. Whales have no hands with which to hold or work material and so could never build a raft. Chimpanzees do not build nets because their survival is not dependent upon technology apart from using sticks to capture termites or leaves to drink water. One might conceive of elephants building small structures out of logs for shade since they have a trunk which functions in many ways like a hand but which is a very poor substitute for one. Still, elephants do not build anything, because like chimpanzees, their survival strategy does not include a reliance on technology.

Our hands, even more so than our brains, have directed the course of technological development. From crude stone tools to lunar landers, our tools and equipment were designed to suit the abilities and limitations of our hands. Hands have had a role in the development of everything which is unique about our species. Without them, we would not be human. Human hands can be elegant and beautiful or powerful and cruel. They are the defining feature of our species, and the source of our technology and culture. We are the Handy Man.

CHAPTER 11

Wise One

The human brain is surely one of the marvels of evolution. At its core is the primitive neural control mechanism of our ancestors going back in time through the mammals, reptiles, and fish. As much as 90% of our brain is required for body maintenance, information gathering from the senses, memory, and motor control. Since the brain evolved gradually, with components added millions and hundreds of millions of years apart, it is highly compartmentalized. This can be seen more readily in fish and amphibians where prominent lobes for smell, vision, muscular coordination and other activities are clearly visible. The large increase in the size of the human brain hides many of the individual components, but they are there none the less. Each area of the brain has a specialized function to perform. The wonder is that despite the highly compartmentalized nature of the brain, all of the areas coordinate and share information. When examining something new, for example, we may touch it, smell it, shake it, and examine it visually. Although this sensory information is all going to different parts of the brain, the information is quickly put together to form a mental construct of the object as well as conjectures about its function, origin, and other matters of interest.

In measuring brain size relative measures are used which compare the size of the brain to the overall size of the body. The first definitive increase in the size of the brain in the human lineage occurred about 2 million years ago when the first members of the genus *Homo* appeared. Lucy's brain was not significantly larger than that of today's chimpanzees, but with the appearance of Handy Man the brain showed a significant increase in size. A second large increase came about 700,000 years ago with *Homo erectus* who had a brain nearly double the size of Lucy's. The largest increase came with the transition from *Homo erectus* to *Homo neaderthalensis* and *Homo sapiens* between 500,000-100,000 years ago. It is significant to note that we are the inheritors of a brain that was completely formed 100,000 years ago, a brain which evolved to deal with environmental challenges much different than we face today.

It is also important to note that the increase in size of the brain was not an overall increase. Older, more primitive areas remained unchanged or were reduced to make way for increases in other areas. Compared to other primates those areas associated with motor skills, smelling, and sight have been reduced, while most of the increase in size came in the frontal areas and occipital lobes of the brain. This increase began even before Handy Man and continued until the appearance of our own species, and it represented a radical change in function from olfactory analysis (smell) to abstract processing. It is the increase in the frontal lobes that gives us our high foreheads relative to our ancestors and which makes possible technological, abstract, and computational thinking.

The human brain burns fuel faster than a high performance jet aircraft

Brain size increases came at a high cost in energy consumption. The amount of energy available to the brain in any animal is a function of how much energy is left over after organ systems like digestion, respiration, skeleton, muscles, etc. are satisfied. This in turn is dependent upon the quality and quantity of food intake. The energy requirement of the human brain is several times that of a chimpanzees brain even allowing for the difference in body size. Approximately 25% of the energy during resting metabolism goes to feed the human brain compared with about 8% in a chimpanzee. Although the brain makes up only 2% of our body weight, it consumes 15% of the oxygen and 40 % of blood glucose.

This raises several important questions. Why would natural selection have favored an increase in brain size which enormously increased the body's demand for food? Gathering food must have been a difficult and dangerous activity for our early ancestors. Not only were they vulnerable to the many large predators roaming the savannah, but they had to compete with dozens of other animals for food. Secondly, how did our hominid ancestors manage to increase their caloric intake to make the increase in brain size possible?

The second question seems to be easier to answer than the first. There is compelling evidence that by the time of Handy Man our ancestors had begun to supplement their diets with increasing amounts of meat. Other living primates feed primarily on fruits, leaves and other vegetable foods. Chimpanzees do occasionally kill and eat monkeys, and they apparently are very fond of meat, but the bulk of their diet comes from plant foods.

Handy Man apparently scavenged food from the kills of other predators and may have hunted small animals himself. He made stone tools which had a sharp edge on them which could have been used to butcher animals. Experiments with these crude tools today indicate that they were very effective in quickly dismembering animal carcasses and scraping the meat from the bones. Many of the fossil bones of prey animals found at Handy Man living sites show characteristic cut marks from these tools. Try to imagine, if you will, getting enough meat off a dead zebra or other animal with only your teeth and fingernails. Handy Man was not much better off than we would be in that situation. Getting through the tough skin would have been nearly impossible without tools of some sort, and even if it were possible, it is likely that predators like hyenas would be attracted to the carcass long before anything of value could be gotten off of it. With sharpened stones, a carcass could be dismembered and carried off quickly. Other bones found at these living sites were broken in a way that indicates Handy man also cracked open and ate the calorie rich marrow of long bones such as those in the legs of large animals like water buffalos and giraffes. These bones are too large and tough for other predators to crack with their teeth and they would have been abandoned at kill sites where they could have provided a ready supply of high quality food for Handy Man.

It is very unlikely that Handy man was able to kill anything other than very small animals. In all probability, he supplemented his diet of tubers, fruits and other plant foods with meat scavenged from animals killed by predators. Meat is easily digested and very high in caloric content. A shift toward meat eating would have made it possible for our early ancestors to acquire the additional food energy required to support an increase in brain size.

I will increase your labor and your groaning, and in labour you shall bear children.
Genesis 3: 16

One of the most dangerous trips any of us will ever take is through our mother's birth canal. Walking upright required that the legs move closer together at the top so that stability on one leg could be maintained during walking and running. This was done by narrowing the pelvis, which reduced the size of the ring of bone the birth canal passes through at the lower end of the pelvis. The newborn of chimpanzee and other apes are in an advanced state of development compared to human babies. Their nueuromuscular development is such that they can cling to their mother's fur as she moves through the trees. A chimp's brain at birth is half the size of an adult's, while the brain of a human child is only a fourth that of an adult. Human babies are completely helpless at birth and require constant care. If human development followed the pattern of chimpanzees and other apes the gestation period would be 21 months, but the size of the brain of a child nearly two years old would make birth impossible because of the narrow birth canal.

Females-helping-other-females during birth is a human behavior which is not found in other primates. Apes have large bodies and a wide pelvis, which makes childbirth relatively easy compared to humans. The structure of the female ape's pelvis also allows the ape infant to be born facing the mother so that she can help it emerge. Human birth is made especially difficult and painful due to the fact that while the baby's head is nearly round, the birth canal varies in width along its course, which means that the baby must twist to fit through the narrow canal. This causes the infant to emerge facing away from the mother. If a human mother tries to help the infant during birth she will be bending it backward and could severely injure its back and spinal cord. The difficulty of the passage through the birth canal is often evidenced by a skull that appears badly misshapen following delivery.

It is impossible to say for sure at what point in human evolution the birth of helpless infants began, but given the brain size of *Homo erectus* it seems very likely that they gave birth to underdeveloped infants. From the neck down *Erectus* was almost identical to a modern human, including the size of the birth canal. In fact, some fossil specimens seem to indicate that the birth canal was smaller in *Erectus* than in modern humans. Measurements of the size of the birth canal in fossil specimens indicate that their children were born much like our own, which suggests that midwifery may have been common among this species. The difficulty of childbirth and the helpless condition of the newborn led to an increase in sociality among early hominids. Women helping during childbirth and the added dependence of new mothers upon others because of the helplessness of the infant required a more complex social structure than is found in chimpanzee and other ape societies. The need for help with the infant may have been the origin of monogamy, which is unique to humans among primates.

Death In the Desert

The extreme difficulty of human childbirth is illustrated by an experience related by a doctor on the western American frontier during the 1880's. The doctor was called to assist at a birth but arrived too late. The woman's husband had tried to assist her, but the results were fatal. Dr. Coe described the scene when he arrived:

> On the bare simmering sand near the water hole, quailing under the hot mid-day sun, stood a tent. In front on the hot sand lay the body of an almost naked woman smeared with sand and blood from head to feet. Grouped around it in a circle were eight ragged, dirty children, from three to eighteen years old, crying and wailing in the most abject misery and grief. A tall, powerfully built man, wild-eyed, ragged and dirty, with a three weeks' growth of beard, the wide brim of his sombrero flopping in time with his movements, was doing the Piute War Dance around the children. He held a naked unwashed baby in his left arm and was brandishing a big six-shooter at the sky with his right hand. In a vituperative stream of blood-curdling profanity he threatened all the gods in heaven and defied Jesus Christ to come down to earth in person and fight him in mortal combat......It was a pathetic, ghastly and soul-sickening sight.
>
> - American Heritage. April 1963. p.91

Humans are the only primates who form monogamous relationships, and the only ones which engage in sex year round. All other primates without exception have a yearly estrous cycle during which females will mate with many males. Males only rarely fight over females and for the rest of the year there is no sexual contact between males and females. The difficulty of human birth has changed all of that. The helpless infant and its mother need the help of others, and this is probably the origin of family structure and year round sex as a method of bonding between a male and female. It could also be the origin of one of the most powerful and deadly human emotions – jealousy. Statistically, jealously is the third most common motive for murder. Hardly a week goes by without the news media carrying the story of a boyfriend killing a woman (and often her children) over a failed relationship or because of infidelity.

Children of God: Children of Earth

Committed: © United Feature Syndicate, Inc.

Erect Man moved out of Africa and into Asia and Europe sometime between 1.5 and 2.0 million years ago. He carried fire with him to warm his nearly naked body and tools for hunting, digging, butchering, and making clothes. Like us, he was a naked ape. The loss of hair in our ancestors is related to the growth of the brain. The enormous amount of energy which the brain consumes generates a great deal of heat which must be eliminated from the body. Heat is carried away from the brain by blood flow to the skin. The heat is then transferred to the environment. This is facilitated by sweating. It takes a great deal of heat to evaporate water, so spreading body fluids over the skin and allowing them to evaporate carries heat away much faster than radiation or convection can do. Sweating is so efficient that heat can be transferred from the body when the environmental temperature is higher than body temperature. A person can survive temperatures considerably higher than body temperature as long as they are sweating. If sweating stops during exertion on a hot day body temperature will rise sharply and death may occur within minutes.

Keeping the brain cool requires many more sweat glands than are possessed by apes, and because hair holds moisture and heat, we have become naked apes with an abundance of sweat glands all over the body. Running, which is one of the main purposes of upright walking, can quickly double the heat production of the body. Having left the cool forests to live on the hot open savannah, natural selection favored the loss of hair and an increase in the number of sweat glands. Increases in the size of the brain drove our evolution even further in that direction. As Erect Man moved into Asia and Europe, the loss of hair necessitated the development of clothing and the tools for making them.

Anthropologists disagree about the selection pressures that led to the growth of the brain and the development of the hand, but one thing is certain, without hands our big brain is useless: it could not feed us or protect us, nor could it have done so for our progenitors. As the hand and brain became more human in the course of evolution the quality and variety of tools produced became greater.

Given the extraordinary costs of a big brain in terms of energy consumption, the dangers of childbirth, and the need to care for a helpless infant, we must ask what benefits could have been so great as to outweigh the costs. In other words, why did natural selection drive human evolution in the direction of the big brain?

One obvious answer, it seems to me, is that moving from the relative safety of the forest to the more dangerous savannahs and grasslands of Africa demanded a more cohesive, integrated social life than the loose shifting relationships of most primate groups. An individual alone, whether it be Lucy, Handy Man, Erect Man, modern man or some other hominid, could not survive for long in that environment. Packs of wild dogs and hyenas, not to mention leopards, lions, water buffaloes and many other dangerous animals would make the survival of a single hominoid impossible. Despite these dangers, food was widely scattered and difficult to obtain, forcing our ancestral species to travel widely and often. In that environment, and under those circumstances, one hominid is no hominid.

The most difficult aspect of life is dealing with other people – spouses, parents, children, coworkers, teachers, salesmen, policemen, neighbors, salesmen, relatives, service people, friends, etc. Doing so successfully requires a good memory for remembering the names of individuals, their behavioral traits, and our past experiences with them so that we can manipulate or at least get along with them. Highly developed problem solving skills are also necessary to deal with the problems created by the unpredictability of social interactions. When a person has had a "bad" day, it is usually because of problems with other people that were not solved satisfactorily. Every one of us has two basic social needs: the need to be part of a group; and, the need to stand out in some way from the group. Achieving these needs in a highly competitive social setting requires guile, highly abstract reasoning, problem solving skills, and good communication skills.

Given the need to act in concert and move and live in tightly knit social groups for survival, our hominid ancestors would have been under strong selection pressure to develop the brain power to deal successfully with the increased social pressures they were functioning under.

Another seemingly obvious answer to why natural selection would favor an increase in brain size and complexity despite the drawbacks is that early tool use, although extremely crude, proved to be so successful on the open grasslands that natural selection placed a premium on technological inventiveness and abstract thinking in order to produce more efficient tools. In part this involved adjustments in the anatomy of the hand and increased brain control over hand movements and use. Sitting in a cave or other home site making a tool out of a shapeless stone requires abstract thinking at many levels. The selection of the stone is critical: is it hard enough to do the job but soft enough to be worked? Is it the right size and weight for the work it will be required to do? The tool maker must conceptualize in his mind how the tool will be used, how it will respond to pressure and wear, and in making decisions about the tool he must draw upon his knowledge and past experience. More importantly, perhaps, he must rely on the accumulated knowledge of generations of tool makers, each of whom may have made some small improvement in design and material selection.

The technological revolution of the twentieth century has demonstrated the power of human technological inventiveness. In less than a century we went from the first airplane flight to standing on the moon and sending probes to other planets, and we did so with a brain that reached its present level of development long before what we call civilization began. During that same century the power of the atom was unlocked and used for both peaceful and violent means. The electronic revolution put a computer in every classroom and most homes, and telephone calls from across the globe became as common as people walking the streets with a cell phone to their ear. But as fantastic as these achievements are, none of them has had as great an impact on our survival as a species as some of the inventions of our ancient ancestors. We lived quite well without airplanes, space vehicles, computers, and cell phones, and although it may seem heresy to our children, had none of these things been invented and developed we would still be living a good life.

The really important inventions, the ones which meant that our species would live or become extinct, are all in the past. The greatest of these, of course, are agriculture, fire, and the domestication of animals, but an apparently simple invention like the spear thrower (atlatl) – nothing more than a short stick cupped at the end so that the butt of a flexible spear would fit into it – was infinitely more important in the survival of our species than all of the inventions of the twentieth century. With the leverage given by the thrower, the spear could be thrown several times farther and with greater power. This simple device, which a person could carve from bone or wood in less than an hour, allowed larger animals to be hunted and permitted the hunter to strike from a much safer distance. Using the atlatl a man could throw a spear twice the length of a football field and kill animals the size of antelopes at fifty yards. At a time when our ancestors were struggling to compete with other predators for food the atlatl may well have meant the difference between prospering and living on the edge of starvation. Hunting is dependent upon cooperation and planning, and as dependence upon a diet of meat grew, there were strong selection pressures for increased brain power which could produce better weapons and tools and which could cope with the increased social pressures which resulted from the highly cooperative nature of group living.

Our survival as a species was by no means ever assured. Studies of human DNA indicate that on more than one occasion population levels dropped to near extinction before rebounding. For most of our history as a species we numbered no more than 5 million at any given time. It is only in the past ten thousand years that numbers have risen as a result of agriculture and other technological innovations. The development of an upright posture, creative hands, language, and a larger brain were all efforts to meet the challenges of a constantly changing and hostile environment, and they all came with severe costs, all of which still bear upon us today. We are space age creatures living in bodies which were designed for a life style that no longer exists except in tiny remote corners of the world. Having mastered the natural world so completely that we are in the process of destroying it, we have yet to learn to master ourselves. Our resourceful and combative ancestors successfully met the challenges of their lives and in so doing passed the torch of life to us. It remains to be seen whether the genetic heritage which sustained them on the plains of Africa and the glaciated lands of Europe contains sufficient variability for us to adapt to a space age world with a population more than a thousand times greater than theirs. To my mind, the issue is in doubt.

CHAPTER 12

Language of the Hands

One invention stands out above all others in defining our humanity – language. Language is both an evolutionary and a cultural innovation, one whose origins are lost forever in the shadows of the past. We are not alone in communicating with each other: all living things, from bacteria and protozoans to humans, do so in one way or another. But no other species has developed a spoken and written language that can convey ideas and emotions with force and subtlety. In an earlier chapter I made the point that hands are what make us human, that everything from our culture to our big brain have their evolutionary roots in our hands, and suggested that the most appropriate name for our species is *Homo habilis*, Handy Man. It would not be inappropriate to amend that title to Handy Man Who Speaks.

Our hands were intimately involved in the development of speech and the relationship continues to this day. Gesturing with the hands is as much a part of speech as producing words and sentences. The hands provide emphasis and subtle meaning to the words we speak. When our early anthropoid ancestors moved into the open savannah they encountered some of the world's largest and fiercest predators. The danger of becoming prey to these animals was very real because long distances had to be covered daily in order to obtain food in an environment where the typical food of primates – leaves and fruits – was scarce. They adjusted by scavenging on the kills of other animals, and by gathering roots and other plant foods, but the loud raucous vocalizations used by apes in the forest could attract predators and get them killed.

It seems very likely that much of their communication while moving about was by gestures, particularly with the hands, in much the same way that soldiers use hand signals when in close proximity to the enemy. Other primates use body language in addition to guttural sounds but they use their forelegs for walking and climbing which restricts their use for signaling. Standing upright freed the arms and hands for multiple uses, including signaling. Hand signals can be used over a considerable distance in open country and are easily understood. Two fingers held upright against the side of the head can indicate an antelope. Waving the hand up and down like the ocean waves can indicate that it is running, and a finger can point the direction the animal is moving. More than that, however, hand movements can express emotion and nuance. Human language may have evolved from manual gestures which survive today as a "behavioral fossil" coupled to speech.

Primitive gesturing can lead to a kind of symbolic representation that is coded. That is, agreed upon hand movements can become a language using coded signals that only those familiar with the code recognize. Two fingers held to the side of the head is a non-coded signal that anyone can recognize as a horned animal, but a thumbs up or down has no obvious meaning except the meaning given it by those using it. It could mean success or failure; I'm hungry; your wife is pregnant; an enemy is coming, or any number of other meanings. It was the transition from non-coded gestures to coded gestures which marked the beginning of human language.

Words are as abstract as a thumbs up or down. They have no meaning of their own; their meaning relies on common agreement among the users. There is nothing in the way a word is written or said that in any way indicates its meaning. At some point in our evolutionary history coded hand gestures and abstract vocalizations began to work together in a way that increased the communicative value of each. This apparently drove natural selection to make changes in the vocal tract and brain which led to improvements in the ability to speak.

Sign languages are today recognized to have all the characteristics of a true language. "There are countless different sign languages invented by deaf people all over the world, and there is little doubt that they are genuine languages with fully developed grammars. The spontaneous emergence of sign languages among deaf communities everywhere confirms that gestural communication is as "natural" to the human condition as is spoken language. Indeed, children exposed from an early age only to sign language go through the same basic stages of language acquisition as children learning to speak, including a stage when they "babble" silently in sign!" (American Scientist Online – the Gestural Origins of language 2/2/2006).

In the 1980's the Nicaraguan government opened a school to educate deaf children. When children were brought together who had no formal training in sign language they created a new language that their teachers could not at first interpret. The language was complete with grammar and syntax and had no oral component. The language developed into more complex forms as newer students added to and refined the gestures of the older students. Learning a language clearly has a genetic component which makes it all but inevitable that children will learn to speak, but the experience of the Nicaraguan students indicates that the process is also dependent upon social interaction and feedback.

The strong selection pressures for survival on the open savannah required a higher degree of sociality than the loose, constantly changing nature of primate groups like those of chimpanzees. Increased sociality and strong group identity would in turn have required better vocal communication than the emotional grunts and screams other primates use. These selection pressures led to changes in the throat and cerebral cortex that over time made human speech possible.

A small, horseshoe-shaped bone in the throat of apes and humans called the hyoid is used to support some of the muscles of the throat. The position of the hyoid bone and larynx (voice box) in apes is high up in the neck. This causes a tight connection between the trachea and nasal cavity which prevents food and liquids from entering the airways and chocking the animal, but also prevents them from speaking as humans do. Interestingly enough, the hyoid and larynx in a newborn child is also high up in the neck. This permits infants to breathe through the nose and swallow liquids at the same time. Gradually, as the infant grows the hyoid and larynx move lower in the neck and by the age of three months have reached the lower position found in adults.

The low position of the hyoid and larynx makes human speech possible, but at a huge cost. Everyone has experienced the terror of choking on food. One has to assume that the risk of suffocation from food or liquid entering the trachea was outweighed in the course of evolution by the increased survival benefits of speech.

Alterations in the cerebral cortex of the brain were also necessary for human speech to develop. A small area on the lower left side of the frontal lobe called Broca's Area, is associated with both speech and hand and arm movements. A stroke to this area leaves a person without the ability to speak. A stroke in Broca's Area in deaf people makes it impossible for them to sign.

A second area in the temporal lobe, called Wernicke's area is also important in the ability to understand words. When this area is damaged there is a memory loss which results in an inability to understand what is being said. Deaf people with damage to this area no longer comprehend sign. In almost all right handed persons these two speech areas are in the left hemisphere of the brain, while in some left handed people they are in the right side. The intimate kinship between hands and speech in human evolution is further confirmed by this relationship.

James R. Curry

People who have lost or never had the senses of sight and sound are much more aware of how much the hands tell us about our world. A young woman who is deafblind wrote the following poem to express the importance of hands in her life.

MY HANDS

My hands are . . .
My Ears, My Eyes, My Voice . . .
My Heart.
They express my desires, my needs
They are the light
that guides me through the darkness
They are free now
No longer bound
to a hearing-sighted world
They are free
They gently guide me
With my hands I sing
Sing loud enough for the deaf to hear
Sing bright enough for the blind to see
They are my freedom
from a dark silent world
They are my window to life
Through them I can truly see and hear
I can experience the sun
against the blue sky
The joy of music and laughter
The softness of a gentle rain
The roughness of a dog's tongue
They are my key to the world
My Ears, My Eyes, My voice…
My Heart
They are me

By Amanda Stine.

Human language did not emerge suddenly in evolution. Because the soft tissue of the throat and brain do not fossilize, we may never know just when the human lineage finally achieved the ability to speak as we do today. A Neanderthal skeleton found in a cave in northern Israel in 1983 contained a complete hyoid bone, the only one ever found with anthropoid fossil remains. The bone indicates that Neanderthal was capable of speech, and it seems likely that *Homo erectus* was also able to speak.

Thinking with words allows for abstract thought and creativity of a kind probably not possible for any other creature. But thinking in words is in some ways extremely crude because it slows thought processes to the speed of a glacier. It has long been debated whether we can have thoughts without words. I believe that when we are faced with a life threatening situation that requires quick action, we revert to a kind of thought process that is much faster because it does not involve words. An incident I once experienced may illustrate.

As a young man with a powerful car I often could not resist taking risks on the highway. Once, caught behind a long line of slow moving vehicles on a two lane country road, I dropped my "four on the floor" transmission into a lower gear and roared out into the passing lane. As I reached the middle of the line of cars I saw a truck coming toward me at a high rate of speed. Already high on adrenalin from the risk I was taking, my adrenal system instantly kicked into high gear. One by one, with a clarity I still remember to this day, I considered my options: brake hard and try to get back to my starting point behind the line of cars – rejected; no time. Brake hard and swerve off to the side of the roadway – rejected; that would take me down a steep slope and onto railroad tracks a hundred feet below. Force my way between two of the cars I was attempting to pass – rejected; the cars were too close together and I was moving past them too fast. Besides, if I forced them off the road a terrible accident involving many cars would result. Drop the transmission into a lower gear and press the accelerator to the floor. Question – is there time? Judging the speed of the truck coming at me, my now frightening speed, and the distance to the end of the line of cars I was passing, I chose this option. It was probably the closest I have ever come to dying in an automobile accident. I got by the line of cars and swerved back into my lane as the truck blasted by me with his horn blowing. It was so close that I can still remember what the front bumper of that truck looked like as it shot by my open window.

Several things stand out about this incident. There was no fear at all, just the mental consideration of the problem and my physical reaction to it. More incredible to me was that running through each of the options and selecting one of them took only a split second of time. I marveled at this when thinking about it later. It has taken me hundreds of words to describe what went through my mind, but I did it all in a split second. Clearly, I was not thinking in words.

I am not the only person to have experienced this kind of thinking. A military pilot once related to me an almost identical experience when he saw another plane coming directly at him. He used the words "time warp" to describe that sense in which time seemed to stand still while he mentally processed the situation and came to a decision on how to act. In fact, he indicated that only a split second of time was involved, just as in my experience, but that it seemed much, much longer. Many of you reading this will have had similar experiences and understand what I am saying, but it will be difficult for those of you without such an experience.

The way language is structured forces us to think in a step-wise, mechanical manner which can be very cumbersome. Other animals do not think in words. There is much controversy today as to whether other creatures are sentient or not. Do they have a conscious awareness of self? Is such a thing possible without language? We can't answer these questions and may never be able to, but my reversion to a nonverbal method of problem solving at a moment of extreme crisis leads me to believe that this is how other animals think and solve the problems of survival. Being unable to learn indirectly from the experiences of others of their kind as we can do through oral and written language, they are unable to prepare ahead for novel situations and must rely on instinctive reactions or rapid nonverbal thought. We have retained that capacity for use in a crisis, despite our almost complete dependence upon language in our everyday lives.

The Peacock's Tail

Charles Darwin suggested that intelligence and imagination evolved in humans through sexual competition for mates. In his book, The Mating Mind, Geoffrey Miller points out that "Every one of our ancestors managed not just to live for a while, but to convince at least one sexual partner to have enough sex to produce offspring. Those proto-humans that did not attract sexual interest did not become our ancestors, no matter how good they were at surviving....it is clear that our minds evolved largely through the sexual choices our ancestors made." Miller goes on to say, "The human mind's most impressive abilities are like the peacock's tail: they are courtship tools evolved to attract and entertain sexual partners." In other words, language, music, art, and creativity in general are the result of women choosing to mate with men who had minds lively enough to entertain them. Miller sees these attractive behaviors as "courtship ornaments" similar to the peacocks tail.

The human mind evolved as a result of the sexual choices our ancestors made?

This overly simplistic approach to the evolution of the human brain ignores a number of obvious problems. Without question female choice is a powerful influence on the direction of evolution, and like the peacock's tail the human brain is extremely expensive to grow and maintain, but focusing only on sexual selection ignores or depreciates the role of natural selection. Creativity is valued most in animals that live a life style that is opportunistic. Plant eating animals that are surrounded by more food than they can consume tend not to be as inquisitive and creative as animals that live by their "wits." A green iguana can sit in the same tree all day feeding on leaves and be perfectly well fed and safe. A predator, on the other hand, must leave its den or home site and search the environment for its food. It must be inquisitive and creative to successfully find and kill its prey while avoiding being killed by other predators. Animals with highly specialized life styles like the iguana make excellent pets because they are usually dull-witted and do not need much in the way of stimulation or challenge. Given enough food they are content to sit alone in a cage for long periods of time. Placing an opportunistic animal like a predator in a cage, on the other hand, is almost a death sentence.

When vegetarian apes moved from the tropical forest, where they fed safely on leaves and fruit, to the savannah, they had to become far more opportunistic in order to feed themselves and avoid being killed. They also had to develop a more cohesive social structure. Both of these necessities undoubtedly played a major role in the development of a larger brain. The most difficult aspect of life is maintaining successful relationships with other people. Mel Gibson's film, *What Women Want,* in which he gains the ability to read women's minds, illustrates how easy it would be to get along with and manipulate other people if we knew what they were thinking. The signals we give each other regarding our thoughts and emotions are good, but subtle, and we are always left guessing. If we could read each other's minds, we would lose the most challenging aspect of our lives.

Many biologists interpret sexual selection in light of the "handicap" principle. A peacock's tail or a moose's antlers, while ridiculous ornaments which endanger the animals, provide a kind of truth in advertising. In effect, a male is demonstrating that he is healthy enough to be able to grow and maintain these expensive structures, and what's more, he can survive even though they are a severe handicap to him. In a sense they are like the ridiculously large and ornate homes in Beverly Hills and elsewhere that flaunt material success. Miller's contention that the human mind is primarily or wholly due to female mate selection in no way satisfies the handicap principle. Clearly, the larger brain, while costly, was of enormous advantage in terms of overall survival. Unlike the peacock's tail, the larger brain was not a handicap to everyday success at obtaining food and avoiding predation.

"Speed Bump" © Dave Coverly/Dist. By Creators Syndicate

We don't know enough about the social structure of early hominids to know what choices females made in the selection of mates or even if they were free to choose mates. Based upon the need to care for children and to have a mate that could feed and protect them, one might assume that hunting ability, status within the tribe, tool-making ability, and general health might have been more important in choosing a male than trinkets or the ability to amuse.

Using words as abstract symbols is a uniquely human activity today, one made possible initially through the ability of the brain to communicate with the hands and the hands to communicate with the brain. No one knows just when or how in our evolution language originated, but it was an event that was to impact every other living creature on Earth. It transformed our species from an insignificant ape to a creator of almost god-like status and a destroyer of worlds.

Children of God: Children of Earth

NON SEQUITUR © 2008 Wiley Miller. Dist. BY UNIVERSAL PRESS SYNDICATE. Reprinted with permission. All rights reserved.

CHAPTER 13

Cain and Abel

A small child stumbled into camp one cold winter day. She was naked except for a tattered loin cloth, and the pallor in her face and swollen belly were clear signs of starvation. She cringed in fear as the men approached her with their weapons at hand. Pitifully, she held out her hand for food, but they gave her none. One of the men grabbed her by the back of the throat and threw her to the frozen ground. The terror in the little ones eyes softened the heart of one of the men, and he called for his wife to help the girl, but the other men would not allow her to approach. An argument arose between the men as to what they should do with the girl. It grew very heated, and in the confusion the little girl snatched a piece of bone lying on the floor of the cave and began to gnaw on it ravenously. The woman took advantage of the quarrel among the men to take the girl aside, but as quickly as it started the quarrel ended when the shaman silenced the others with a wave of his staff and some strong words. Without any further comment, he snatched the girl away from the woman, grabbed her by the ankles, and swung her violently against the rock wall. She died instantly, and her small body was thrown outside for the animals to fight over. With her died a species of humankind, Neanderthal, for she was the last of that species. The silence that followed Neanderthal's extinction persists to this day, for it was the first of tens of thousands of human-related extinctions which were to come, but more than that, it was Cain killing his brother Abel. Now only one human species was left to inherit the Earth.

The first fossils identified as Neanderthal man were found in a cave in the Neander Valley near Dusseldorf, Germany, in 1856. Recognized as being human, the fossils were named *Homo sapiens neanderthalensis*. Since then, other fossils have been discovered in France, Spain, Italy, Yugoslavia, China, Java, and

Israel. The fossils indicate that Neanderthal man inhabited Europe and western Asia from about 200,000 years to 30,000 years ago. The climate in these regions was much colder than it is today, and several ice ages occurred during the time of Neanderthal's occupation.

Neanderthal man has always gotten an undeservedly bad press. Portrayed as an animalistic brute, he was in fact very human in every respect. His large nose, short muscular build, short limbs, and thick heavy bones were adaptations for living in cold climates, adaptations similar to those seen by people living in sub-arctic regions today. Neanderthal's average brain size was slightly larger than modern man's (1450cc).

Neanderthals constructed complex shelters, used fire, wore clothing, and buried their dead. They were formidable hunters of large animals and their stone tools and weapons were similar to those of the modern humans living at the time. They were apparently capable of speech, as evidenced by the discovery of a fossil hyoid bone in a cave in Israel, and hollowed out bones with holes drilled in them may indicate that they made musical instruments similar to flutes. Body ornaments made of shells have also been found associated with their living sites and they made representatives of animals in bone and other materials, perhaps indicating that they practiced a form of religion. Despite the obvious similarities in the life style and capabilities of Neanderthal, many people continue to think of them as dumb brutes. They were in fact highly intelligent people with the capacity for abstract thought, and they were sufficiently adaptable to survive in the cold, hostile regions of Europe during several ice ages.

Reconstruction of a Neanderthal child from Gibralter (Anthropological Institute of Zurich)

In 1868, at Cro-Magnon rock shelter in the village of Les Eyzies in

France, five skeletons were found which had been buried along with stone tools, carved reindeer antlers, and pendants made of shells and ivory. Cro-Magnons were, for all practical purposes, modern humans. Their brain size was slightly larger than modern humans, but they were completely modern in appearance. They built huts, made cave paintings and carvings, wove cloth for clothing, and had ritualized burial of their dead. There is also strong evidence that they cared for their elderly, the sick, and the maimed.

Cro-Magnon are thought to have migrated into Europe about 35,000 years ago, which brought them into contact and competition with the Neanderthals who had been there for some 200,000 years. Archaeological evidence shows that at least initially, the Cro-Magnon people occupied primarily hilltop living sites while Neanderthals occupied the valleys and lowlands. This may indicate that as Cro-Magnon moved into the region they were forced to occupy sites not already occupied by Neanderthal, or it may indicate that they sought out the hilltops for their better defensive position.

Competition between the two peoples was inevitable as they hunted the same food and used the same raw materials. The archaeological record indicates that neither had an advantage over the other in hunting skills or the weapons they used for hunting. The disappearance of Neanderthal around 29,000 years ago is an enigma which is controversial today among scientists. One commonly held theory is that the two peoples merged through interbreeding, giving rise to modern people. This theory has been weakened, if not destroyed, by the results of recent studies in Germany of DNA taken from Neanderthal bones. The studies showed wider genetic differences between Neanderthal and modern man than was previously supposed. The differences are great enough to suggest that Neanderthal was a distinct species from modern man. The findings show that Neanderthal mitochondrial DNA differs by 7% from living humans. By comparison, chimpanzee chromosomal DNA differs from humans by less than 2%. As a result of these studies many scientists today prefer to refer to Neanderthal as *Homo neanderthalensis*, rather than *Homo sapiens neanderthalensis*. The genetic differences may indicate that successful interbreeding between the two species was not possible.

Still, one wonders if some of the Neanderthal genes did not make their way into modern man. Given the similarities in appearance and behavior, rape seems a likely possibility between the two groups, as does child stealing. American Indians, for example, often took children from the settlers and raised them as their own. Even if rape and child-stealing did occur, however, the two peoples may not have been genetically close enough to make crossbreeding successful.

So, the disappearance of Neanderthal from the Earth after hundreds of thousands of years of successful existence remains a mystery. Although superficially different in appearance, he was in many ways similar to the races of modern man which existed at the time. All had a tool technology, hunted animals, used fire, wore clothing, spoke a language, and made representations of the things around them.

It seems likely, given the intolerance which we see today between races, religious groups, and peoples of different cultures, that Cro-Magnon and Neanderthal would not have been able to coexist peacefully. Quite possibly, then, Neanderthal was eliminated as a species through warfare between the two groups, perhaps the earliest example of genocide on a large scale. Human groups throughout history have been extremely protective of their living and hunting areas and have only rarely been willing to cooperate with other groups or share lands or resources. The similarities in life styles made the two peoples extreme competitors for land, food and natural resources.

It is also possible that when Cro-Magnon moved into Europe he brought with him a disease or diseases to which Neanderthal had no resistance. The introduction of new diseases into a population in olden times was often catastrophic. It is estimated, for example, that 90% of all North American Indians were killed by diseases like small pox and measles which were introduced by European settlers. Viruses and pathogenic bacteria mutate frequently, producing new strains which can originate in a small number of people and then sweep through an entire population. The Spanish Flu of 1918 which circled the globe and eventually infected 20% of the world population began with a few soldiers in Fort Riley, Kansas. Twenty to forty million people died as a result of the flu, a high percentage of who were healthy young people. If this did occur to Neanderthal, it would not have wiped out the entire population, but might have made a takeover by Cro-Magnon less difficult.

The story at the beginning of this chapter of the little Neanderthal girl is fiction, but in some way and in some time, the last Neanderthal died. Perhaps warring with modern humans, which must have been extremely vicious and bloody, is partly responsible for the way we war against each other today. Internecine warfare between these two groups of very intelligent and well armed humans would surely have created a strong selection pressure for aggression, technology, organization, group identity, specialization, technological innovation, and all the characteristics which are the hallmarks of our culture, and which we apply so freely to warring against each other today. The meek would certainly not have inherited the earth following such a prolonged confrontation, a confrontation which could have lasted for centuries, if not longer.

Is it possible that Cain killing Abel is a reference to modern man killing his brother, Neanderthal? The people living when the last Neanderthals were slain would surely have incorporated such a momentous event into their oral traditions, and as we know from modern tribes, oral traditions can be extraordinarily long lived. One can imagine the gray-bearded elders sitting around the campfire late at night telling the youngsters of their tribe about a race of people that used to inhabit the land and how they were driven from it by the children's ancestors, tales which could have later become incorporated into the book of Genesis.

If, as is beginning to seem likely, Neanderthal was a different species of humanity, the Judeo-Christian dogma of our centrality as God's only children is wiped out in a single stroke. Why would God create two species of humans, only to let one of them destroy the other?

Science Purloined

Scientists' have no control over how their work is used (or misused). Efforts to purloin science through ignorance of its precepts or to justify or further political, economic, or religious agendas have repeatedly led to attacks on the scientific community. Scientists themselves have sometimes been guilty of misapplying their theories in ways that brought condemnation from their colleagues and the public. The next three chapters deal with two examples of science purloined: the eugenics movement in the United States and Europe and the ongoing attacks by conservative religious groups on the teaching of evolutionary theory in the United States.

CHAPTER 14

Three Generations of Imbecils

Carrie Buck was a poor, unwed 17-year old mother living in Charlottesville, Virginia, when she was chosen to test a 1924 Virginia law which allowed state institutions to surgically sterilize "inmates of state institutions who are insane, idiotic, imbecile, feebleminded, and epileptic, and who by the laws of heredity are the probable potential parents of socially inadequate offspring." Carrie's mother had been committed to a state institution for the Epileptic and feebleminded and Carrie was originally put in the care of foster parents and later transferred to a state institution.

The state of Virginia claimed that Carrie was unfit to reproduce because she was promiscuous and "feebleminded." The superintendent of the state institution in Lynchburg where Carrie was involuntarily confined testified against her saying, "These people belong to the shiftless, ignorant, and worthless class of anti-social whites of the South."

The case eventually went to the Supreme Court in Buck v. Bell. Doctors who had examined Carrie's baby testified that the child was "below average, and "not quite normal." Justice Oliver Wendell Holmes rendered the formal opinion in the case on 2 May, 1927:

"It is better for all the world, if instead of waiting to execute degenerate offspring for crime or to let them starve for their imbecility, society can prevent those who are manifestly unfit from continuing their kind... Three generations of imbeciles are enough."

Carrie was sterilized, and a precedent was set which upheld a state's right to involuntarily sterilize people in state institutions deemed to be genetically defective. It was subsequently shown that Carrie was not promiscuous, feebleminded, or the mother of a retarded child. Her pregnancy was the result of rape by a relative of her foster parents, and her daughter was a solid B student in first grade and received an A in deportment (She died of illness at age 8). It was also later revealed that Carrie's attorney conspired with the state against her, and at least one of the doctors who testified during her trial had not even examined her.

Before the Virginia law was repealed in 1974, 8,000 people were involuntarily sterilized, and nationwide the number rose to almost 70,000, the majority of whom were poor women who were unable to stand up for themselves against the power of the state. In Europe the number rose to almost half a million.

Eugenics is the self-direction of Human evolution

The above logo is from the 2nd International Congress of Eugenics in 1921. The word Eugenics was coined by Francis Dalton, an English geneticist. It is derived from the Greek and means "well-born." The movement to improve human genetic qualities through compulsory sterilization of the "unfit" became popular in the United States and some European countries during the early part of the 20th century.

The first sterilization bill in the United States was introduced in the Michigan legislature in 1897, but was defeated. In 1899 Dr. Harry Sharp began sterilizing inmates of the Indiana State Reformatory in Jeffersonville. He eventually sterilized over 220 men before it became legal. A bill aimed at "Idiots and imbecile children" was passed by the Pennsylvania legislature in 1905 but was vetoed by the governor. Indiana passed the first sterilization law in 1907, and eventually 30 other states and 2 Canadian provinces followed suit. Most of these laws were not repealed until the mid 1980's. The North Carolina law was not repealed until 2003, although the Sterilization Commission which oversaw sterilizations was abolished in 1977. In Europe, Sweden continued compulsory sterilization until the 1970's.

Sterilization represented only a small part of the agenda of those supporting the eugenics movement in the United States. Eugenicist's supported and fought for bans on interracial marriage, restrictions on immigration from non-European countries, and limited welfare for the poor.

The American eugenics movement was initiated largely through the efforts of Biologist Charles Davenport, Head of the Eugenics Record Office (ERO) based at Cold Springs Harbor Laboratory in New York. Davenport wrote eugenics textbooks which were widely used in high schools and colleges. In 1911 he appointed Harry Laughlin, a colleague at ERO, to head up a research program to determine the national origins of "hereditary defectives" in American prisons, mental hospitals, and other charitable institutions in order to influence the ongoing Congressional debate on curbing the growing tide of immigrants. Laughlin published a "Model Eugenical Sterilization Law" in 1914 which became the basis for the Virginia law as well as for laws in other states.

Laughlin's model law authorized sterilization of the "socially inadequate", "feebleminded", "insane", "criminalistic", "epileptic", "inebriate", "diseased", "blind", "deaf", "deformed"; and "dependent – including orphans, ne'er do wells, tramps, the homeless and paupers". It is ironic that Laughlin, an epileptic, could have been sterilized under his own proposed law.

Laughlin was influential in Nazi Germany during the 1930's and was awarded an Honorary Degree in 1936 from the University of Heidelberg for his work in behalf of the "science of racial cleansing." His law greatly influenced similar laws passed by the Nazis under which nearly 400,000 people were sterilized. Although the Nazi laws were originally directed at similar health or social problems as were targeted in the United States, these were later used to attack whole populations of Jews, Gypsies, Poles, and other people judged by the regime to represent "worthless lives." When the Nazi administrators of the sterilization programs went on trial at Nuremberg, they justified their actions by citing the United States as their inspiration.

The eugenics movement in the United States was primarily a political movement by an economic, social, and racial elite which sought to maintain racial, class, religious, and sexual dominance. In 1907, the year Indiana passed the first compulsory sterilization act, immigration to the U.S. passed 1 ¼ million, up from 150,000 only twenty years earlier. As the sheer numbers of immigrants grew, eugenicists allied themselves with other interest groups to restrict immigration. Eugenicists were also alarmed at what they perceived as a decline in the birthrate of the "better classes" coupled with the surging birthrates of immigrants, poor native-born, and the "unfit." Also of concern, in the words of a leading California eugenicist, was the "evils of (racial) crossbreeding."

A well known Sacramento banker, Charles M. Goethe, one of the founders and sponsors of the Eugenics Society of Northern California, stated in 1929 that the Mexican is "eugenically as low powered as the Negro…He not only does not understand health rules: being a superstitious savage, he resists them." Goethe stumped the state promoting efforts to curb immigration from below the border and sterilize the "socially unfit".

Support for the eugenics movement came from all sectors of society, but in retrospect it was primarily dominated by white, upper class, Protestant, men, and its racial overtones are undeniable. In its heyday in the 1920's and

1930's schoolchildren and college students studied from biology textbooks which contained a chapter on eugenics, movies promoted it, and county fairs all over the country had eugenics exhibits and held competitions to determine outstanding familial blood lines. Awards were given to families whose blood lines were considered to be superior. Eugenics doctrine was even preached from some pulpits; in fact, awards were handed out by the American Eugenics Society for particularly noteworthy sermons. Eugenics became a part of everyday life – for those who deemed themselves to be above defect.

The most chilling aspect of the eugenics movement is that the proponents were not obscure eccentrics or fringe extremists; they included some of the most influential and well know personalities from the upper and middle classes. College presidents, Nobel-Prize winning scientists, noted capitalists, rabbis and ministers, politicians, bank presidents, medical doctors, and teachers were included in their membership. There was no segment of society that was not involved in one way or another. President Wilson signed the New Jersey sterilization law, and Theodore Roosevelt commented that "society has no business to permit degenerates to reproduce". Alexander Graham Bell and Margaret Sanger were also supporters of involuntary sterilization.

Some biologists spoke out against what they saw as social injustice and bad science. The Nobel Prize winning geneticist Thomas Hunt Morgan denounced the movement as unscientific, and the noted geneticist Hermann Muller denounced the American eugenics movement as racist, elitist, and sexist at the 1932 meeting of the International Congress of Eugenics.

Eugenics was a form of social Darwinism. It's founder, Francis Dalton, was greatly impressed by Darwin's Origin of Species and thought that the concept of selection could be applied to the human population. The leaders of the eugenics movements in Europe and the United States seized on Darwinian selection to further their political and social agendas. It was not until after World War II, when the atrocities committed by the Nazi's began to come to light, that the eugenics movement in the United States died. To this day, however, the precedent set in Buck v Bell that made it legal to involuntarily sterilize the "feebleminded" under state control, has not been overturned.

A relatively small but influential group of people perverted the sciences of genetics and evolution in an effort to justify and protect the prevailing racial and class hierarchy in the United States, and they were supported in their actions by state legislators and the Supreme Court. Almost seventy thousand people had their reproductive rights taken away as a result.

Although the eugenics movement of the 20th century is dead, much of the social injustice which it sought to perpetuate is still with us in the form of economic, racial, and sexual elitism, not to mention the tragedy of "ethnic cleansing" which we have seen practiced in Africa and the former Yugoslavia in recent years.

The Genetics of Negative Genetics

Negative eugenics – sterilizing people considered as "unfit" – can never achieve the goals set by the proponents and practitioners of the eugenics movement. Harmful human traits are usually inherited as recessive genes, but some are also inherited as dominant genes, and others are due to a combination of many genes, each of which affects the trait in some quantitative or qualitative way.

The frequency of harmful recessive genes is kept very low in the population naturally. Cystic fibrosis can serve as an example. This devastating inherited disease clogs the lungs with thick, sticky mucus which leads to life threatening lung infections. The disease also prevents the pancreas from producing the natural enzymes which break down food. The gene for cystic fibrosis occurs with a frequency of .02 (20 in 1000) among white North Americans. There are only two forms of the gene, normal (C), and mutant (c), therefore the normal gene occurs with a frequency of .98. For a person to have cystic fibrosis both parents must be carrying the recessive gene: the probability of this happening is approximately .0004 (less than one-half of one percent of all marriages). When both parents are carrying the harmful recessive gene there is a 25% probability that their children will have the disease.

Individuals with cystic fibrosis are carrying two recessive genes for the trait (cc). If every person in the United States who is afflicted with cystic fibrosis could be identified and sterilized, it would take approximately 50 generations (1500 years) in which every person with cystic fibrosis was sterilized to reduce the frequency of the harmful gene from .02 to .01. Since it would be impossible to actually identify and sterilize everyone with the trait, it would undoubtedly take even longer. The reason for the slow decline is that .02% of the population is carrying one copy of the gene (Cc) but does not show its effects because they also have a copy of the normal gene. Clearly, sterilizing every child with cystic fibrosis would have almost no effect in controlling the disease.

Negative genetics is also ineffective in eliminating genetic disorders which result from dominant genes. Dominant genes which are lethal before a person can reproduce can't hide from natural selection the way recessive genes can; consequently, they are already at the lowest level possible in the population. Normal genes, whether dominant or recessive, mutate from one form to the other. This occurs in every generation, largely the result of radiation like ultraviolet light acting upon the genetic material in eggs and sperm cells.

Lethal dominant genes will be reduced in frequency to the extremely low level of mutation rate, and sterilization could not reduce it any further.

Some harmful dominant genes do not act early in life and cannot be detected before the afflicted people have children. Symptoms of Huntington disease, for example, do not begin to appear until around the age of 40. The disease then leads to death within about 15 years. Since Huntington disease is the result of a dominant gene, only one gene can cause the trait. A person with an afflicted parent has a 50% chance of getting the harmful gene. If all children from a parent with the disorder were to be sterilized, by the rules of chance 50% of them would **not** be carrying the harmful gene and sterilizing them would be a moral and legal travesty.

Many genetic disorders are inherited through a series of genes, each of which affects the disorder qualitatively. This allows the disorder to be expressed in a range of conditions from least harmful to most harmful. Intelligence and some forms of mental retardation follow this pattern. Deciding where to draw the line regarding sterilization would of necessity be extremely arbitrary with these traits.

Chromosomal aberrations also cause genetic disorders. Most of these are fatal prior to or shortly after birth, but a few are not. Down syndrome is one of these. It is due to the fact that the child has inherited three #21 chromosomes rather than the normal two. The third chromosome somehow interferes with normal development prior to birth and causes a variety of problems, including some mental retardation. The causes of chromosomal aberrations are not known, but they are more likely to occur in older mothers. Since chromosomal disorders can occur in virtually any woman, no sterilization program could control them.

Quite apart from the ethical, moral, and democratic principles which are violated in sterilizing people against their will, negative eugenics is impractical from a scientific standpoint. It simply will not work with human populations.

Yuppie Eugenics

Today they are desperate to send their kids to the best schools; tomorrow they could be desperate to give them the best genes money can buy.
- Michael Cook

Every scientific breakthrough comes with a high cost in terms of social acceptance and adjustment. Transitions of any kind are difficult, whether they are individual or public. For everything that is gained, something may be lost. The computer revolution, for example, has increased our ability to communicate, order merchandise, pay our bills without leaving home, and keep track of our daily agendas. Computers allow us to find information in seconds that would be difficult or impossible to find by other means, and they offer a unique source of amusement and entertainment. On the negative side, computers threaten our financial identity, cause us to communicate electronically with people we would have formerly walked down the hall to meet face to face, addict our children to mindless games, provide sexual content that is inappropriate for any but the most depraved, and make it possible for our children to form alliances with people who can do them harm.

Scientific breakthroughs can also cause social upheaval on a massive scale, as evidenced by the revolution and controversy attendant to Darwin's theory of Natural Selection which continues in the United States almost 150 years after the publication of his book. Today we are experiencing a biotechnology revolution that is expanding its control over our food supply, medical and drug industry, personal health and reproduction, and which many people fear will expand its influence into what Ruth Hubbard has called Yuppie Eugenics. Yuppie Eugenics differs from the eugenics of a century ago in that it is a matter of individual choice; of wealthy parents genetically engineering their children to certain specifications.

Are designer babies desirable? A neighbor of mine recently had his eight year old son held back a year in school so that he would be taller than the other boys in his class and have a better chance of making the basketball team. Another man I am acquainted with sold his house, gave up his job, and moved to another town so that his son could play basketball under a better coach. Would people like this hesitate to use genetic engineering to emphasize "good genes", in this case genes for athleticism, in their children?

There are severe technical problems in bringing about designer babies. Traits like athleticism, musical and artistic ability, physical beauty, and intelligence are controlled not by individual genes but by hundreds of thousands of genes. Athleticism, for example, depends upon height, muscular strength and coordination, eye-hand coordination, a competitive spirit, intelligence: in short, it is a whole-body potential. It may be possible to genetically engineer one of these traits like height, but in the process others may be diminished.

Nonetheless, scientists like Professor Lee Silver of Princeton University think that genetically engineered humans are inevitable. He foresees a class of people, the "GenRich" who have synthetic genes for health, intelligence, longevity, and athletic ability. Says Professor Silver, "The GenRich class and the Natural class will become…entirely separate species with no ability to cross-breed, and with as much interest in each other as a current human would have for a chimpanzee." Fortunately for all of us, such technology is still a long way from reality. And not everybody has the same values as Professor Silver.

Silver is correct in his assumption that genetic engineering could lead to castes within the human population. If we can genetically engineer better athletes, musicians, artists, brilliant scientists, why not also engineer hyper aggressive soldiers, dull-minded assembly line workers, and extremely small people for space flight?

Biotechnology does hold out promise for positive eugenics – the elimination of "bad genes" through genetic engineering, but only on an individual basis, not for entire populations of people. Eliminating genetic disorders like cystic fibrosis is certainly worth pursuing. The mutation responsible for cystic fibrosis was discovered by Dr. Francis Collins and a colleague in 1989, and it is possible to sample the DNA of a fetus to determine if it carries the gene, but the gene cannot be repaired by today's technology. That will come in time, but in the meantime the only options for the parents are to abort the child or allow it to be born knowing it will suffer a potentially deadly disease.

There exist today a small but growing number of proponents for a eugenics program involving genetic engineering. James Watson, coauthor of the now famous 1953 paper on the structure and function of DNA made the following statement in 2005 during an interview on the BBC: "If you are really stupid, I would call that a disease. The lower 10 percent who really have difficulty, even in elementary school, what's the cause of it? A lot of people would like to say, 'Well, poverty, things like that.' It probably isn't. So I'd like to get rid of that, to help the lower 10 per cent."

In 2007 Watson was quoted in The Times of London as suggesting that overall people of African descent are not as intelligent as people of European descent. He later issued an apology for the remarks stating that "there is no scientific basis for such a belief." Within days of making the comments he was relieved of all administrative duties associated with his position as Chancellor of the prestigious Cold Spring Harbor Laboratory on Long Island. He subsequently resigned that position.

Watson's statements reek of the same arrogance, elitism, intolerance, and social discrimination as the credos of the eugenicists of a century ago, but Watson is not alone in that kind of thinking. In a world where millions of children starve to death annually and the majority of the world's population lives in poverty, Yuppie Eugenics seems like a narcissistic preoccupation of overindulgent Americans who grow more obese every day and who are killing themselves with overwork in order to maintain a lifestyle based upon material success and exclusivity.

"Yuppie Eugenics" also has its boosters among journalists and professional bio- ethicists. Michael Kinsley, writing in *Slate* (April 2000), suggested that genetic tests should eventually be used as qualifications for employment. He was seconded by Andrew Sullivan in the *New York Times Magazine* (July 2000), where he argued that genetic testing for future capacities is less objectionable than using SAT scores or letters of recommendation, since genetic tests are "more reliable." Arthur Caplan, Glenn McGee, and David Magnus of the University of Pennsylvania Institute of Bioethics follow the Kinsley-Sullivan thesis to its logical conclusion when they state, in a 1999 *British Medical Journal* article, "it is not clear that it is any less ethical to allow parents to pick the eye color of their child or to try to create a fetus with a propensity for mathematics than it is to permit them to teach their children the values of a particular religion or require them to play the piano." www.zmag.org/Zmag/articles/march02hubbard-newman.htm

One has to wonder what the result will be of allowing people to pick and choose between particular traits in their children. And if we go down that road, what is lost? Will we lose resistance to disease or artistic ability? Will we create children who have so little in common with each other that they cannot form healthy relationships? Do we lose the genetic diversity that has allowed our species to survive for a million years or more? Do we give up our democratic ideals and practices? Do we lose sight of what it means to be a worthy human being or to respect the differences of others? Do we lose our sense of community and humanity? Genetic recombination unencumbered has allowed the human race to survive and prosper. Short circuiting the process or attempting to control it may create children without the broad range of aptitudes and resilience that have been a hallmark of our species.

Perhaps the day will come when a baby not yet born will have its DNA analyzed to determine its IQ, proclivity for disease, physical features, potential for athletics, the arts, science, and other traits; information which will be available later in the child's life (on line?) to employers, insurance companies, schools, the military, and anyone else with a "need to know." The child's life will be plotted out at birth based upon the genetic analysis. Which schools he will attend, which careers he can pursue, which level of society he may enter, whether he can get health and life insurance – all may depend upon the genetic screening prior to birth. Because only the wealthy will be able to afford the technology to create the "GenRich", poor people will be kept down forever and a class society will develop which will effectively destroy any semblance of democracy. People who propose moving society in that direction obviously feel that they and their descendents will always surface to the top in a class-stratified society. Any geneticist could tell them otherwise.

James R. Curry

Even with the help of technology, genetic recombination will work its wonder in building and tearing down highly successful combinations of genes.

> *Eugenic and other gene dreams will not cure what ails us.*
> - Ruth Hubbard and Stuart Newman

CHAPTER 15

Barefoot Scientist

Science and politics sometimes become curious bedfellows. Such is the case of Trofim Denisovich Lysenko who became the embodiment of the mythical Soviet peasant genius, and who became a virtual dictator in the biological sciences under Stalin. Born to peasant parents in the Ukraine, Lysenko held Soviet genetics captive from the late 1920's until the early 1960's. Rejecting Mendelian genetics and the chromosome theory of inheritance, he believed that environmental influences on plants could alter them genetically in a way that increased their ability to survive. In order to develop strains of wheat that were cold hardy, for example, he subjected seeds to extreme cold and believed that this altered them genetically into cold hardy plants. Biologists had long since shown that this "Lamarkian" concept of inheritance could not be demonstrated scientifically, but Lysenko believed that a "practical" approach to scientific research was preferable to a theoretical one. Consequently, he rarely followed up his work with the kind of careful research that western geneticists were conducting, and often used false documentation to support his work.

COULD WE EVER HAVE DREAMED OF SUCH A GREAT HONOR?
Letter from Academician T.D. Lysenko's Parents to Comrade Stalin

Our beloved, dear Stalin! The day we learned that our Trofim was awarded the Order of Lenin was the most joyous day of our lives. How could we ever have dreamed of such a great honor, we, poor peasants from the village of Karlovka in Kharkov province?

It was hard for our son Trofin to get an education before the revolution. He was not admitted – a peasant boy, a muzhik's son – into the agronomy school, even though he received only the highest grades in school. Trofim had to become a gardener in Poltava. He would have remained a gardener for life had it not been for the Soviet regime. Not only the older Trofim, but his younger (brothers and sister) went to study at institutes... The high road to knowledge was opened up to the muzik's son... is there any other country in the world where the son of a poor peasant could become an academician? No!...

> With kolkhoz greetings!
> Denis Lysenko (father of Academician Lysenko) and Oksana Lysenko (mother). Odessa, January 2.

"Pravda," 3 January, 1936. From: Valery N. Soyfer, "Lysenko and the Tragedy of Soviet Science," Rutgers University Press, 1994; p. 73.

The Soviet leadership supported Lysenko and gave him almost absolute power over Soviet science because of his influence with the peasant population. The peasants viewed him as one of their own, and the resistance that peasant farmers showed toward Soviet practices declined dramatically under Lysenko's leadership. He was put in charge of the Academy of Agricultural Sciences of the Soviet Union by Stalin and directed to take action against scientists whose ideas threatened the party line. Lysenko did this with a vengeance, causing hundreds of scientists to be expelled, imprisoned, or killed. In the process he also destroyed the once flourishing field of Soviet genetics. He was directly responsible for the death at the hands of the NKVD of Nikolai Vaviliv, one of the Soviet Union's most prominent biologists.

Lysenko's brutal control of Soviet Science was weakened in 1962 when three prominent Soviet physicists spoke out against him. In 1964 Andrei Sakharov destroyed Lysenko for good when spoke out against him in the General Assembly of the Academy of Sciences:

> *"He is responsible for the shameful backwardness of Soviet biology and of genetics in particular, for the dissemination of pseudoscientific views, for adventurism, for the degradation of learning, and for the defamation, firing, arrest, even death, of many genuine scientists."*

Soviet genetics and agriculture were greatly retarded as a result of Lysenkoism. The lesson to be learned is that political control of science and scientists are antithetical to the free inquiry which is necessary to drive science forward. While western geneticists were creating a "Green Revolution" with new hybrid strains of wheat and rice derived through genetics and Darwinian selection, Soviet farmers were struggling to feed themselves. Economically and scientifically Lysenkoism was a disaster for the Soviet Union.

Using science in inappropriate ways to advance political agendas is dangerous to society and to individual freedom. Today, in the United States, the religious right is aggressively pushing a political agenda much broader than Creation Science or Intelligent Design. What they want is to change the tempo of society, to eliminate a woman's right to abortion and to legislate against homosexuality and other behaviors they feel are immoral. As Michael Shermer points out in his book, *Why People Believe Weird Things*:

> *"..as soon as a group sets itself up as the final moral arbiter of other people's actions, especially when its members believe they have discovered absolute standards of right and wrong, it marks the beginning of the end of tolerance and thus reason and rationality. It is this characteristic more than any other that makes a cult, a religion, a nation, or any other group dangerous to individual freedom."*

The line between necessary government regulation and stifling control over scientific research must not be crossed if science is to flourish. It is true that scientists are not always as responsible as they should be in their quest to test their theories, and some governmental regulation is necessary. I cringe visibly, for example, when I think of Enrico Fermi and his team of physicists building and putting into operation the first atomic pile under the football stadium at the University of Chicago. None of those men were completely sure that an atomic reaction of the kind they were initiating could be controlled. Fortunately, they did control it and the atomic age was born, but if the reaction in their primitive atomic pile had gotten out of control, an explosion would have occurred which could have destroyed the university and part of the city.

Basic scientific inquiry in the United States today is largely controlled by governmental agencies and multinational corporations through their funding of research. Scientific research is expensive, and universities generally will not keep a research scientist on unless he or she attracts an impressive amount of research money. Corporations also hire scientists to produce new products or refine older ones, and governmental agencies like NASA are directly involved in scientific research. The field of genetics, in particular, has become big business through the production of new products like medicines and genetically modified foods, and new discoveries, including genetically modified organisms, are regularly patented.

Scientific inquiry and technological innovation proceed at breakneck speed today, dragging society and culture behind them. We should not accept this without question. The quality of our lives and the lives of the other billions of people who share the planet with us are at stake. Perhaps we should pause to consider the kind of world we have created and where we want to go in the future.

...and the cost of a thing is the amount of what I will call life which is required to be exchanged for it, immediately or in the long run.
- H.D. Thoreau

CHAPTER 16

Creationism and Intelligent Design

When I was a child, I spake as a child, I understood as a child, I thought as a child: but when I became a man, I put away childish things.

1 Corinthians 13:11

The story of Noah and his Ark are known to almost every school child. God was angry because of the wickedness of man, and He decided to destroy the Earth and all that was in it. Only one man, Noah, found favor in the eyes of the Lord. Noah was commanded by the Lord to build a boat (ark) and to bring into it seven pairs, male and female, of all clean beasts and fowl, and two pairs of unclean beasts so that they might replenish the Earth following the deluge the Lord had planned. Once this was accomplished, it began to rain and continued for forty days and forty nights. Noah was 600 years old at the time. The entire Earth, including the highest mountains, was inundated and every living thing except the animals on the ark and Noah's family was destroyed.

As the flood water began to subside the ark came to rest atop the mountains of Ararat. When the waters had subsided completely, Noah and his family and all the creatures left the ark and were commanded by God to "Be fruitful, and multiply, and replenish the earth."

Most Archeologists, historians, and Biblical scholars regard the Genesis account of Noah's Ark as a very ancient story handed down by oral tradition prior to being recorded in the Bible. A Mesopotamian account of a flood found in the Epic of Gilgamesh is remarkably similar to the Biblical account, although it is not as old.

Accepting this story as literally true rather than as allegory is difficult for any rationally thinking person for a number of reasons, the most compelling one being that there is absolutely no evidence whatsoever that a world-wide flood of this magnitude ever occurred. After literally hundreds of archeological excavations in the Near East no stratum of flood debris has ever been found. Neither is there any physical evidence that people ever lived to be 600 hundred years old; indeed, very few ancient people lived beyond the age of thirty.

Biologists can only estimate how many different species of animals live on the Earth today, but the estimates range from 10-60 million. Despite centuries of collecting and studying animals from all over the planet by thousands of naturalists and biologists, no accurate figure for how many species exist is yet available. To believe that one man, Noah, working alone or with his family, was able to gather together at least two of every species in a relatively short period of time simply defies rationality.

A flood of the type described in Genesis would have destroyed all land creatures, and most of the marine and freshwater animals as well. Most animals that live in salt water die if placed in fresh water. Some exceptions, however, can be found among animals living in estuaries where they are repeatedly subjected to changes in the salinity of the water. Likewise, animals that live in freshwater usually die when placed in salt water. So, unless Noah had huge aquaria aboard the Ark, and somehow managed to collect and transport all of the creatures from the oceans and freshwater lakes and rivers, these waters would be almost barren of life today.

People who accept a literal interpretation of the flood story also believe that the Earth is only six thousand years old. For a tiny number of animals to have repopulated the entire Earth from a single spot in the Middle East in less than 6,000 years is biologically impossible. When the Biblical scriptures were written humans had not yet developed the concept of deep time. At best, the ancients could conceive of two or three generations of the past based upon living relatives or stories handed down. The ability to conceptualize time in millions or billions of years lay centuries into the future of the Biblical writers. To literally accept their story on the age of the Earth is as illogical as it would be to limit ourselves to their technology and methods of agriculture.

It is also difficult to accept that one man and his family constructed a boat of the size indicated in the Bible. They would have had to cut down hundreds of trees, move them to the place of construction, saw them up into lumber, and manhandle huge beams into place several stories above ground level. The impossibility of this, even with God's instructions, seems obvious.

Numerous expeditions to Mt. Ararat like the ones conducted by former astronaut James Irwin have combed the mountain but found no credible evidence of the Ark. So much about the story of Noah and his Ark defy rational thinking that I am confounded that so many otherwise rational people accept it as literally true. And yet, similar flood stories are so widespread among different cultures that one has to believe that it is based upon a regional flood (or perhaps a tsunami) of such magnitude that the story was passed down orally for thousands of years before Biblical writers described it. In many respects the Bible is an accurate record of the events of antiquity, however, the power in the Bible lies not in its historical accuracy but in its guidance and reassurance. People of conservative faith defend the literal description of the flood because they believe that the Bible is the word of God, and that every word in it must be accepted literally. In other words, they accept the story of Noah on faith, and faith by definition does not require evidence.

NON SEQUITUR 2008 ©Wiley Miller. Dist. BY UNIVERSAL PRESS SYNDICATE. Reprinted with permission. All rights reserved.

Given that faith requires no evidence, how do we understand the efforts of the modern Creationist movement in the United States to use pseudo-scientific jargon as evidence for the Biblical stories of creation and the flood? Consider the following abstract of a paper delivered at the third International Conference on Creationism, held in Pittsburgh in July, 1994:

"Any comprehensive model for earth history consistent with the data from the Scriptures must account for the massive tectonic changes associated with the Genesis Flood. These tectonic changes include significant vertical motions of the continental surfaces to allow for the deposition of up to many thousands of meters of fossil-bearing sediments, lateral displacements of the continental blocks themselves by thousands of kilometers, formation of all of the present day ocean floor basement rocks by igneous processes, and isostatic adjustments after the catastrophe that produced today's Himalayas, Alps, Rockies, and Andes. This paper uses 3-D numerical modeling in spherical geometry of the earth's mantle and lithosphere to demonstrate that rapid plate tectonics driven by runaway subduction of the pre-Flood ocean floor is able to account for this unique pattern of large-scale tectonic change and to do so within the Biblical time frame."

Using the language of science, and sounding as if it is based upon solid scientific methods of research and experimentation, it is in fact nothing of the kind. What it is, is an attempt to prove "scientifically" that the flood occurred in the manner and time described in the Bible, that as a result of the flood the land features of the earth today – the mountain ranges, oceans, lakes, plains, valleys, etc. - were produced by the flood in about a year's time, and that this happened within the last ten thousand years. The problem is that the study is not based upon any scientific research nor does it present any evidence to back up the claims made by its author. It is, in fact, mere supposition.

Contrast this with a scientific study of many years duration conducted by two senior scientists from Columbia University who have found evidence of a great flood in the area of the Black Sea about 7500 years ago. William Ryan and Walter Pitman have theorized that during the ice age the Black Sea was an isolated freshwater lake surrounded by farms and small villages. As the earth began to warm about 12,000 years ago the vast sheets of ice which covered the northern hemisphere began to melt, raising sea levels. The water in the Mediterranean Sea rose behind a natural land barrier that separated it from the much lower fresh water Black Sea.

The land barrier gave way suddenly, as a dirt dam will do when water begins to flow over the top. The water roared through a narrow opening into the fresh water lake for at least 300 days at two hundred times the force of Niagara Falls. The level of the lake rose rapidly, flooding 60,000 square miles of land and converting the fresh-water landlocked lake into a salt water lake with access to the world's oceans. Naturally in such a rapid deluge, the farms and villages which surrounded the lake were destroyed. So much water flowed through current day Turkey into the Black Sea that the level of the world's oceans dropped by a foot. For comparison, try to image what it would be like if the Pacific Ocean could suddenly find an opening into Death Valley. A wall of water nearly three hundred feet high would sweep away everything before it and scour the valley down to bedrock.

Sand dunes originally created by the wind have been discovered by cameras in the Black Sea at 140+ feet depth, and joint studies by Pittman and Ryan with a Russian marine exploration team using sophisticated seismic equipment have revealed a uniform layer of marine sediments overlaying sediments obviously formed in a terrestrial environment by rain, wind, and running water. Core samples taken along the length and breadth of the floor of the Black Sea confirmed the seismic studies. Clays in the cores showed clear evidence of mud cracks caused by drying out. The cracks were filled with wind-blown sand and thus preserved. Many of the core samples contained the remains of woody plants, grasses, and other land plants in the material below the marine sediments. Bivalve shells taken from below the marine sediments were of fresh water species, while those taken in higher sediments were of marine species. Radiocarbon dating of the shells confirmed a date of 7,000 years. A 2002 expedition to the Black Sea by Dr. Robert Ballard, the discoverer of the Titanic, revealed an ancient shoreline under the Black Sea which was several miles off today's shore and hundreds of feet deep. Along the ancient shoreline he found evidence of human habitation, including wood structures, artifacts, and a trash heap.

Does the work of Ryan, Pittman, and Ballard prove that the Black Sea deluge is the source of the Genesis story of a world-wide flood? No, but it is far more compelling than a literal interpretation of a story which seems so obviously to have originated in an actual event, but which was converted into a morality story through countless generations of oral transmission.

> *There is more to truth than the facts.*
> - Observed on a Church Sign

Why should it matter to people of faith whether the story of Noah is a literal description of an actual historical event or simply a beautiful morality fable which untold generations of ancient people used to illustrate their concept of the man-god relationship? The answer, I suppose, is that some people cannot think in relative terms but must hold to absolutist thinking. They see their unquestioning acceptance of the Bible as literal truth to be a sign of great faith: I see it as a weakness of faith that is stultifying. We are among the world's most inquisitive creatures, always searching, testing, exploring, changing, challenging. It is a part of our nature that was formed on the open plains of Africa long before modern man appeared, when survival required that every opportunity, no matter how small or unlikely, be investigated for some means of exploitation. To deny this aspect of our nature is to deny our humanity. Absolutism is not faith, it is anti-faith. True faith is not turned to stone by conflicting or controversial details.

I am reminded of a man, a personal friend of my father's, who was disliked by his coworkers for his exemplary work habits, so they conspired to spread rumors that the man's wife was being unfaithful to him. As the stories spread they became more and more explicit and lurid. My father was close enough to the man to bring up the subject with him, which he did one day. "Oh yes, I've heard what people are saying," he replied, "but I know my wife loves me and would never do anything like that." He said this with an air of quiet assurance that left no room for doubt. "Why do you let them go on talking like that without saying anything?" my father asked. The man thought for a moment before answering. "Why? I know what they are saying is not true. That's enough for me." In other words, the man's faith in his wife did not need to be defended to anyone but himself. He understood that the more fuss he made in defense of his wife, the more his faith in her would be brought into question.

I feel the same way about religious faith. It needs no defense, and the more fuss that is made in defending one's faith, the more it is brought into question. Faith is the substance of things hoped for, the evidence of things not seen. To search the mountains of Ararat for years looking for physical evidence of an ark is to question ones faith in my opinion. This goes likewise for perverting science (and theology) in order to prove that the Earth is only 6,000 years old or that theories like evolution must be incorrect because some people believe that they disagree with a literal interpretation of Genesis.

Natural theology is not a new concept. Up until the time of Darwin scientists almost without exception believed that nature was God's creation. This was the basis of their science. In the eighteenth century William Paley likened the universe to a watch which, in his reasoning, shows clear evidence of having been deliberately created by an intelligent being. Darwin himself believed strongly in natural theology prior to his voyage aboard the Beagle.

Creationism was resurrected in the 1960's in the United States by Henry Morris, a Virginia Tech engineering professor. In order to show that science proves religion, he made scientific sounding arguments for a six day creation of the world. None of these arguments were based upon scientific research. A young Californian named Tim La Haye became a disciple and created the Henry Morris Institute for Creation Research. During the 1970's creationism was embraced by the Christian right, after which they became very active politically in statehouses and school boards in an attempt to change the science curriculum to include creationist doctrine. As a result of their activities several states and numerous local school boards mandated that equal time be given to teaching creationism along with evolution.

Arkansas had a 1928 statute on the books which made it unlawful for a teacher in any state-supported school or university "to teach the theory or doctrine that mankind ascended or descended from a lower order of animals," or "to adopt or use in any institution a textbook that teaches" this theory. Any teacher violating the provisions of the act was to be dismissed from his position. Susan Epperson, a young biology teacher in the Little Rock school system, challenged the statue in the State Chancery Court in 1965. The State Court found in her favor, holding that the statute was an abridgement of free speech violating the first and fourteenth amendments. This ruling was overturned on appeal by the State Supreme court in a two sentence opinion that upheld the state's right to specify public school curriculum. The case was then accepted by the U.S. Supreme Court (Epperson v Arkansas, 1968) which ruled that the statute was unconstitutional. "Mr. Justus Fortas delivered the opinion of the court.

> *"the law must be stricken because of its conflict with the constitutional prohibition of state laws respecting an establishment of religion or prohibiting the free exercise thereof. The overriding fact is that Arkansas' law selects from the body of knowledge a particular segment which it proscribes for the sole reason that it is deemed to conflict with a particular religious doctrine; that is, with a particular interpretation of the Book of Genesis by a particular religious group. The antecedents of today's decision are many and unmistakable. They are rooted in the foundation soil of our Nation. They are fundamental to freedom."*

During the early 1980's the Louisiana legislature passed a law entitled the "Balanced Treatment for Creation-Science and Evolution-Science in Public School Instruction Act." The act required that creationism be taught along with evolution. The law was passed after aggressive lobbying by creationists. The lower courts ruled against the state stating that the State's purpose was to promote the religious doctrine of creation science. The case went to the U.S. Supreme Court (Edwards v Aguillard, 1987) which ruled that:

> *"The act is facially invalid as violative of the Establishment Clause of the First Amendment, because it lacks a clear secular purpose."*

> *"Furthermore, the contention that the act furthers a "basic concept of fairness" by requiring the teaching of all of the evidence on the subject is without merit. Indeed, the act evinces a discriminatory preference for the teaching of creation science and against the teaching of evolution ....." with the "purpose of discrediting evolution by counterbalancing its teaching at every turn with the teaching of creationism."*

> *"The Act impermissibly endorses religion by advancing the religious belief that supernatural being created humankind. The legislative history demonstrates that the term "creation science," as contemplated by the state legislature, embraces this religious teaching. The Act's primary purpose was to change the public school science curriculum to provide persuasive advantage to a particular religious doctrine that*

> *rejects the factual basis of evolution in its entirety. Thus, the Act is designed either to promote the theory of creation science that embodies a particular religious tenet or to prohibit the teaching of a scientific theory disfavored by certain religious sects. In either case, the Act violates the First Amendment."*

The Supreme Court's ruling in Edwards v Aguillard forced creationists to go on the defensive, but not for long. Changing tactics to comply with the Supreme Court's ruling, within two years a creationist textbook, *Of Pandas and People*, was produced which was proclaimed by its authors to be an open, objective evaluation of arguments for and against evolution. The book is really a thinly disguised religious tract promoting creationism in the form of intelligent design. The book was sponsored and copyrighted by the Foundation for Thought and Ethics (FTE) in Richardson, Texas, an organization that promotes creationism. The authors, editors, and critical reviewers listed in the book are not identified by title or profession, but simply listed by name. The book makes no explicit reference to biblical passages, and God is identified only as an "intelligent agent," the designer and creator of all living things. Scientists who have examined the book closely have characterized it as a creationist sham masquerading as science. They point out that it misrepresents the fossil record, ignores the geological time line for the history of the earth, and ignores the extinction of species, all of which contradict creationist teaching.

The FTE has actively promoted "Pandas" for use in public school science classes as a supplemental text and insists that it be placed in school libraries as a science text. Reviews by scientists describing the errors and misrepresentations in the book have appeared in many publications, including Scientific American (July 1995, Science and the Citizen, "Darwin Denied"). Comments by Kenneth R. Miller, Professor of Biology at Brown University, are representative of the almost universal reaction of scientists to the book: "Pandas miss-states evolutionary theory, skims over the enormous wealth of the fossil record, and ignores the sophistication of radiometric dating. How sad it would be, given the need to improve the content and rigor of science instruction in this country, for this book to be offered as part of the educational solution. The most compelling reason to keep this book out of the biology classroom is that it is bad science, pure and simple." The publication of "Pandas" by a creationist organization, and the aggressive push by creationists all over the country to get the book accepted as a valid science textbook, shows how far many of the opponents of evolution are willing to go to defeat what they consider to be a challenge to the Bible and their religion. Their misrepresentation of the book as an open, objective evaluation of arguments for and against evolution when it is really a thinly veiled religious effort to discredit the theory makes their efforts seem desperate and beneath the dignity of self-proclaimed conservative Christians.

Mike Keefe, The Denver Post, 2005.

In 2002 the Cobb County school board in Georgia adopted a new science textbook which contained material on evolution. Three parents of students in the school district submitted official comment forms objecting to the use of the book. One of them, Marjorie Rogers, a self-proclaimed six-day biblical creationist, circulated a petition containing 2300 signatures protesting the use of the book. Responding to the complaints, the school board had a sticker placed in the books:

> *"This textbook contains material on evolution. Evolution is a theory, not a fact, regarding the origin of living things. This material should be approached with an open mind, studied carefully, and critically considered."*

Five parents of students brought suit against the board of education in federal court in an effort to have the stickers removed. Among the statements made by Judge Clarence Cooper in his written decision were the following:

> *"The Sticker .. has the effect of implicitly bolstering alternative religious theories of origin by suggesting that evolution is a problematic theory*

even in the field of science. In this regard, the Sticker states, in part, that "[e]volution is a theory, not a fact, concerning the origin of living things" that should be "approached with an open mind, studied carefully, and critically considered." This ..characterization of evolution might be appropriate in other contexts, such as in an elective course on theories of origin or a religious text. However, the evidence in the record and the testimony from witnesses with science backgrounds, including the co-author of one of the textbooks into which the Sticker was placed and Defendants' own witness, Dr. Stickel, reflect that evolution is more than a theory of origin in the context of science. To the contrary, evolution is the dominant scientific theory of origin accepted by the majority of scientists."

"the sticker has already sent a message that the school board agrees with the beliefs of Christian fundamentalists and creationists. The School Board has effectively improperly entangled itself with religion by appearing to take a position. Therefore, the Sticker must be removed from all of the textbooks into which it has been placed."

CHAPTER 17

Return to Mysticism

Our ancient ancestors attributed everything they could not understand in the natural world to spirits. In order to placate or gain the favor of these spirits, they made sacrifices and engaged in elaborate rituals. One by one the spirits have disappeared as science illuminated the dark shadows of our understanding about nature. We are no longer frightened by solar eclipses because we better understand the mechanics of our solar system. We still fear disease, but most of us no longer blame it on spirits (or God) or think that it is a punishment for our sins. Knowledge empowers us by taking away fear and offering hope.

The Holy Grail of science is that natural phenomena have natural causes. In one form or another most people have bought into this concept and believe, perhaps too strongly, that if science can't solve our problems today, it will in the future. When a new disease like AIDS appears, for example, we fully expect scientists to come up with an effective treatment. Today scientists are scanning the solar system for asteroids and comets which might be on a collision course with the Earth. Should one be detected, we would turn to science to divert or destroy it before it could cause a cataclysmic explosion which would destroy us.

Science has been and continues to be an extremely powerful way of understanding the natural world, and scientific knowledge increases at an ever expanding rate, but scientist will be the first to admit that our knowledge is still very limited and our theories open to examination. The certainty exhibited by Christian creationists stands in sharp contrast with the uncertainty scientist's encounter as a regular and natural part of their work, and yet the influence of science in our society is so great that creationist's use scientific sounding jargon to authenticate their religious beliefs.

A case in point is a book published by Michael J. Behe, Professor of Biochemistry at Lehigh University, titled *Darwin's Black Box; The Biochemical Challenge to Evolution*. Unlike most creationists, Behe is a respectable scientist from a reputable institution. Overwhelmed by the molecular complexity of living cells and organisms and by the inability of science to show that such complexity could have evolved under natural conditions on the early Earth, Behe has concluded that life must have had a Creator.

This concept, called Intelligent Design (ID), is not new; it goes back at least to the time of Aristotle. To my mind, it really goes back to the mysticism of the ancients. Simply stated the basic premise of intelligent design is that if science can't explain something today, God must be responsible. Intelligent design means that all forms of life began abruptly through an intelligent agency, rather than through a long period of modification and evolutionary change. Behe has added a new wrinkle to this ancient argument with a concept he calls "irreducible complexity." He starts his argument for ID by quoting Darwin:

> *"If it could be demonstrated that any complex organ existed which could not possibly have been formed by numerous successive, slight modifications, my theory would absolutely break down."*

He goes on to point out that some systems seem to be too well integrated and complex to have formed from less complex systems by successive modification. To make his point he uses a simple analogy; the design and workings of a mouse trap. The trap consists of only five parts, but if any one part is removed, the trap becomes completely nonfunctional. Behe argues that since all parts must function together or the system breaks down, it is impossible to see how it could have developed through small, successive modifications of some prior system. In other words, a mousetrap is irreducibly complex.

The flaw in this reasoning is that a simple mousetrap was in fact created as a unit and did not develop in successive steps through time. The automobile, a much more complex "system" than a mousetrap, did develop from successive small modifications through time. My grandfather once remarked to me that "you could work on a Model T Ford with a pair of pliers and a screwdriver." That is certainly not true of cars today. As a young man I could lift the hood of a car and understand everything I saw in the engine compartment. That, too, is no longer the case. If someone who had never seen an automobile were to lift the hood of a modern car and begin on his own to try to understand how it worked, I feel certain he would be completely overwhelmed by the complexity of the machine. And, like Behe, he would probably feel that the machine is irreducibly complex because removing even the smallest part from the engine or power train causes the car to cease functioning.

If that same person was then given a complete series of old motor cars going back through to the 1890's, he could, through careful examination of them see how the vehicle "evolved" from one which could be worked on with a few simple tools along a backcountry road to one that is so complex it requires specialized mechanics using computers and other highly complex tools to repair.

In those rare instances where the fossil record provides a whole incremental series of fossils through time, biologists are in the same position as the man studying a series of cars from old to modern. The fossil record of horse evolution, for example, shows a long continuous sequence of fossils which document the evolution of a small, dog-sized animal with 4 toes on the front feet and 3 on the hind into the large animal of today with a single "toe" on all feet and the vestigial remains of the other toes still present. Although the development of a modern horse took many twists and turns, the fossil record clearly documents the changes that took place and the timing of those changes.

A research team led by paleontologist Neil Shubin of the University of Chicago and Edward Daeschler of the Philadelphia Academia of Natural Science reported in 2006 the discovery of an evolutionary "missing link" between fish and land animals. Daeschler of Philadelphia's Academy of Natural Sciences recently reported that several fossils of the creature between four and nine feet long were found on Ellesmere Island in northern Canada. The animal had scales, fins, and gills, and was obviously a fish, but it had a moveable head like an alligator and could drag itself along on land. Given the scientific name *Tiktaalik roseae,* the animal lived 375 million years ago in an equatorial river delta before continental drift moved the North American land mass northward. The find is extremely significant because it is an undeniable link between fish and land tetrapods, including amphibians, reptiles, birds, and mammals.

Tiktaalik lived about twelve million years before the first four-footed animals appeared. Tiktaalik shared features of both fish and tetrapods. It had gills, scales, and fins, but its head and body were flat, with eyes on top of the head much like a crocodile. The animals also had a neck, a feature not found in any modern fish. Its ribs were more like those of a tetrapod than fish, which helped support its body and perhaps helped it to breath air. The fins were positioned both for swimming and for lifting its body above the surface. Daeschler has inferred from this unusual combination of traits that Tiktaalik lived in shallow water and used it's powerful front fins to push along the bottom and help it maintain its position against the current.

Bill Day: © Commercial Appeal/Dist. By United Feature Syndicate, Inc.

When the fossil record is lacking or incomplete, as is often the case, biologists have living organisms to study in order to try and unravel the sequence of modifications which may have led to the plants and animals alive today. For example, studying the hands of all the different species of living primates along with the fossil bones of now extinct hominids like Lucy, Handy Man and Erect Man can help us to understand how our own unique hands evolved.

The creationist argument that "if science can't explain it God must have done it" is bogus. There was a time when scientists had no idea what caused disease. During the Black Death in Europe, many people fled to their churches, believing that the disease was God's vengeance for sin and that the church would be a sanctuary against the disease. It was not, and people died as quickly in their cathedrals as they did on the street. There are countless examples of superstition and ignorance giving way to scientific insight in human history. When Benjamin Franklin invented lightening rods many clergy throughout the young country were incensed. To their minds lightening is the wrath of God and should not be tampered with. If it is God's will for a house to be struck, it will be struck, they argued. It is ironic that churches throughout the country soon had lightening rods on them and many were undoubtedly saved from destruction by having them. The value of science in this case was that it empowered people in their fight against lightening fires in a way that their religion did not. The same can be said about Pasteur's germ theory of disease; indeed, it can be said about science in general.

> *Everything God made has a crack in it.*
> - Emerson

Creationists tend to ignore questions for which they have no answer. They ignore extinction, for example. Ninety-nine percent of all the creatures that have ever lived on the Earth are now extinct. Why would God create species only to see them destroyed? Likewise, why would a perfect God create imperfect creatures? The Bible tells us that we are God's most perfect creation, made little lower than the angels. Why then do we have so many imperfections? Why did he place the opening to our lungs in front of the opening to our stomachs so that food could go down "the wrong way?" Did he want us to choke to death? Why did he place the male prostate gland around the urethra? Did he want middle aged and old men to get up several times each night to urinate? Why did he give us a backbone that fails so often? Why are there design flaws in our eyes that limit the amount of light reaching the retina, and flaws which allow the retina to become detached? The squid's eye has none of these problems. Are we to believe that a designer whose greatest creation was mankind made such an incredible error in the construction of the human eye and not in that of the squid? Why do we have an appendix that is essentially a useless organ but which can become infected and kill us?

Are we to assume that the Creator is a malevolent designer, or perhaps just an inept one? These are flaws in design that even a weak engineering student would not make. Evolutionary theory can explain such "flaws": intelligent design cannot. If a Creator had designed and created each and every living thing we would not expect obvious flaws or anomalies (why should blind cave animals have eyes with lenses and retinas, or birds that can't fly have wings?) All of these apparent flaws and anomalies are explainable by evolutionary theory; none are by Intelligent Design.

> *Render therefore unto Caesar the things that are Caesar's*
> *and unto God the things which are God's.*
> - St. Luke 20: 25

On December 14, 2004, eleven parents of students filed suit in federal court against the Dover Area School District in Pennsylvania. The matter at issue was the introduction of "intelligent design" into the school's biology curriculum. The issue began when the school board began consideration of a new biology textbook in June of that year. The book recommended by the teachers was criticized by the chair of the DASB Curriculum Committee as being "laced with Darwinism." The member went on to state that he was looking for a textbook that gave a balanced view between creationism and evolution. Further, he stated that "This country wasn't founded on .... evolution....This country was founded on Christianity, and our students should be taught as such." At a later meeting the same board member is reported to have said, "Two thousand years ago someone died on a cross. Can't someone take a stand for him?"

After much discussion and disagreement on the board it was proposed that the textbook recommended by the teachers be accepted but that *Of Pandas and People* be added as a second textbook. After much acrimonious debate and a tied vote on accepting only the teacher-recommended biology book, one board member changed her vote and the biology textbook passed without *Pandas*. The board members who supported *Pandas* then arranged to have 50 copies of the book donated anonymously to the school district. These were accepted and put into the library and the matter seemed to be closed.

The teachers, however, apparently made it known that they had no intention of making use of Pandas, and this may have been what led the board to list *Pandas* in the curriculum as a reference text and pass the following resolution on October 18:

> "Students will be made aware of gaps/problems in Darwin's Theory and of other theories of evolution including, but not limited to, intelligent design. Note: origins of life will not be taught."

The reference to intelligent design was opposed by the school administration, the head of the science department, and all but one of the 12 citizens testifying at the meeting. On November 19, a press release on the school web site included the following statement which biology teachers would be required to read at the beginning of the evolution unit:

> "The state standards require students to learn about Darwin's Theory of Evolution and to eventually take a standardized test of which evolution is a part.

> "Because Darwin's Theory is a theory, it is still being tested as new evidence is discovered. The Theory is not a fact. Gaps in the Theory exist for which there is no evidence. A theory is defined as a well-tested explanation that unifies a broad range of observations.
>
> "Intelligent Design is an explanation of the origin of life that differs from Darwin's view. The reference book, Of Pandas and People, is available for students to see if they would like to explore this view in an effort to gain an understanding of what Intelligent Design actually involves. As is true with any theory, students are encouraged to keep an open mind.
>
> "The school leaves the discussion of the Origins of Life up to individual students and their families. As a standards-driven district, class instruction focuses on the standards and preparing students to be successful on standards-based assessments."

Eleven parents of students filed suit against the Dover Area School District ID policy in federal court on December 14, 2004. The trial began on September 26 and lasted six weeks. A few days after the close of testimony in the trial Dover voters overwhelming voted eight of the nine board members out of office and replaced them with eight candidates who publicly stated that they were opposed to the teaching of ID in science classes. The ninth member was not up for reelection.

Judge John E. Jones rendered his opinion on December 20, 2005. In it he delivered a stinging attack on the Dover Area School Board, saying that its decision to insert intelligent design into the science curriculum violates the constitutional separation of church and state.

> "The proper application of both the endorsement and Lemon tests to the facts of this case makes it abundantly clear that the Board's ID policy violates the Establishment Clause. In making this determination, we have addressed the seminal question of whether ID is science. We have concluded that it is not, and moreover that ID cannot uncouple itself from its creationist, and thus religious, antecedents.
>
> "Both defendants and many of the leading proponents of ID make a bedrock assumption which is utterly false. Their presupposition is that evolutionary theory is antithetical to a belief in the existence of a supreme being and to religion in general.

Judge Jones further stated that ID "violates the centuries old ground rules of science by invoking and permitting supernatural causation;" it relies on "flawed and illogical" arguments; and its attacks on evolution "have been refuted by the scientific community." He also decried the "breathtaking inanity of the Dover policy and accused several board members of lying to conceal their true motive, which he said was to promote religion.

> INTELLIGENT DESIGN IS NOT SCIENCE
>
> INTELLIGENT DESIGN IS NOT BIOLOGY
>
> INTELLIGENT DESIGN IS NOT AN ACCEPTED
>
> SCIENTIFIC THEORY
>
> Memo From Dover science teachers to the superintendent

The Dover episode was the first attempt by creationists to get ID into science classrooms. When the Supreme Court struck down efforts to introduce creationism into science classes in 1987, its proponents, ever aggressive in defense of a literal interpretation of the flood and creation stories in Genesis, seized on Behe's concepts of ID and irreducible design. He was and is ideally suited to serve their needs for two reasons: he is a respectable scientist, and his ideas dovetail perfectly with theirs. His willingness to ascribe supernatural causes to natural phenomena, and to do so in scientific language, has convinced many people not familiar with either science or the fundamental concepts of ID to propose that it be given equal treatment in public school science classes.

The school board members who testified during the Dover trial, for example admitted under oath that they did not understand the concept of ID. In his written decision Judge Jones said the following:

> *"Defendant's asserted secular purpose of improving science education is belied by the fact that most if not all of the Board members who voted in favor of the biology curriculum change conceded that they still do not know, nor have they ever known, precisely what ID is."*

Earlier in his decision he also said:

> "The Board consulted no scientific materials. The Board contacted no scientists or scientific organizations. The Board failed to consider the views of the District's science teachers. The Board relied solely on legal advice from two organizations with demonstrably religious, cultural, and legal missions, the Discovery Institute and the TMLC."

What the Dover school board had done was the result of the religious views of one or two board members who were able to influence the other members to go along with the proposed change. Two board members who voted against the resolution resigned in protest when it passed. One of them, Casey Brown, said the following during her resignation speech:

> "There has been a slow but steady marginalization of some board members. Our opinions are no longer valued or listened to. Our contributions have been minimized or not acknowledged at all. A measure of that is the fact that I myself have been twice asked within the past year if I was "born again". No one has, nor should have the right, to ask that of a fellow board member. An individual's religious beliefs should have no impact on his or her ability to serve as a school board director, nor should a person's beliefs be used as a yardstick to measure the value of that service.
>
> "However, it has become increasingly evident that it is the direction the board has now chosen to go, holding a certain religious belief is of paramount importance."

These are the methods used by creationists in their attempts to get their religious beliefs introduced into science classes: identify people in public office who agree with them; press them to initiate legislation or decisions in favor of their cause; attack science and scientists as atheistic; and aggressively seek support from conservative Christians, many of whom have no idea what ID stands for. The truth of the matter is that many scientists are people of faith, and that science itself is neutral on the subject of religious faith. The pillaring of science and scientists as aggressively atheistic is particularly effective with conservative Christians who fear that their children will be led away from their faith and family values.

The appeal of concepts like creationism and ID, apart from the religious impetus, is their simplicity. They require no research or factual evidence to back them up, and they are neither provable nor falsifiable. Despite all of Behe's scientific-sounding writings in favor of ID, he has conducted no research into the subject and published no papers dealing with the subject in refereed scientific journals, nor will he because scientific research regarding supernatural phenomena is by definition impossible. Science can only investigate natural phenomena. His ideas are little more than conjecture masquerading as science.

The Indiana State Legislature in 1897 considered "a bill for an act introducing a new mathematical truth, and offered as a contribution to education....by the State of Indiana." The bill proposed to change the value of the mathematical constant pi from 3.14159 to 3.2. This would have made circles and spheres in Indiana larger than anywhere else in the world. The bill passed the House unanimously and it passed its first reading in the Senate before a math professor from Purdue University who just happened to be in town and heard what was happening rushed to the statehouse and explained the ludicrous situation to the legislators. The matter was then dropped.

As ridiculous as this sounds, it is not much different than what happened in the Dover school board. Without consulting their teachers or getting advice from any science authorities, the board took it upon itself to make a substantial change in the science curriculum. They did so even though they did not understand the concept of ID. When school board members were asked why they voted to introduce ID into biology classes many of them said it was because they were told by two of the board members that it was good science. When one of the two members cited by the others was asked by a reporter to give a simple definition of ID, here is what he is reported to have answered:

> *Other than what I expressed, that's — Scientists, a lot of scientists — don't ask me the names. I can't tell you where it came from. A lot of scientists believe that back through time, something, molecules, amoeba, whatever, evolved into the complexities of life we have now.*

Something similar happened in the Indiana legislature in 2007. After meeting in private with a creationist evangelist a number of the leading members of the legislature announced plans to introduce a law requiring public schools to teach ID along with evolution in science classes. No science educators or research scientists were consulted prior to the announcement, and I feel quite certain that the legislators did not understand the concepts of ID any better than the members of the Dover school board. I also feel quite certain that to date the Indiana legislators involved have done nothing to educate themselves about evolution or ID.

Creationist's represent themselves as religious folks who are combating secularism in science and society, and this leads many people to buy into their arguments without any real thought. They project a victim mentality that resonates with religious people who are concerned about the increasingly permissive and secular nature of society. Playing upon these concerns, a tiny minority of people are able to influence the direction of public policy.

Freedom in the Classroom

"Our nation is deeply committed to safeguarding academic freedom, which is of transcendent value to all of us and not merely to the teachers concerned. That freedom is therefore a special concern of the First Amendment, which does not tolerate laws that cast a pall of orthodoxy over the classroom."
- Supreme Court in Keyishiam v Board of Regents (1967).

Academic freedom is the right of teachers at all levels to teach their subjects without undue influence or coercion from the state or other authority figures. This freedom is critical to the functioning of a free society. Dictating for political, religious, or economic reasons what teachers can teach in their own fields is the hallmark of repressive societies. One of the first actions taken by the Nazis after gaining political control of Germany was to rewrite textbooks and publicly burn books that did not agree with their political agenda. We see the same control of education by radical clerics in some Muslim countries today.

If school boards or legislators are permitted to dictate what will be taught in science classes a precedent will be set that allows minority groups of all kinds to impose their particular philosophy or viewpoint onto our educational system. There are people in this country who believe that the holocaust never occurred, that current descriptions of it are the result of a deliberate Jewish conspiracy, and the people who believe this nonsense are supported by a small number of historians. What if a small group of like-minded people gained control of a school board or legislative body? Should they be permitted to force history teachers to give equal time to the belief that the holocaust is a Jewish conspiracy, that there never was an official policy to murder Jews and other people deemed unfit by the Nazi regime, and that homicidal gas chambers did not exist?

Allowing special interest groups to control what is taught can lead to the banning of books from the classroom and library, the teaching of racial inferiority in certain areas of the country, and in the words of the Supreme

Court, "cast a pall of orthodoxy over the classroom."

People who want intelligent design taught as an alternative to evolution say they want young people to be able to choose for themselves what to believe. Nothing could be farther from the truth. In fact, they want children to not believe evolution – that is their real goal, and their agenda is both religious and political. All of our freedoms rely on what is taught in the nation's classrooms, and any attempt to restrict teachers to a religious, racial, or political viewpoint in order to promote a particular agenda is contrary to the best interests of society at large.

> *"Professor Curry, do you want me to answer these test questions by what you taught us in class or by what I know to be true?"*
> – College Student

This was a question put to me by a general biology student after she looked over the questions on a test about a unit on evolution. I thought for a moment, then answered that inasmuch as the exam was designed to test what she had learned in class, she should answer in that way. She said okay, aced the test, and got an A in the class. I have never tried to convert students to evolutionary theory; all I want is for them to understand the basics of it. Too many people who say they are opposed to evolution have no real understanding of the concept. The creationists say that they want ID taught so that students can choose for themselves what to believe, but at the same time they constantly misrepresent evolutionary theory to their followers. This makes efforts of science teachers more difficult, the overwhelming majority of whom have no desire to "convert" students but merely wish to educate them.

I am very sympathetic to people who are concerned that their family values will be diluted or destroyed in the public schools or in college. It has been my experience, however, that most young people who come from families with strong religious and moral values will not turn their backs on their families or their values because of what they learn in school.

One evening as I surfed the TV channels I came across a sermon/lecture by Ken Ham, the Australian-born minister who is the founder and president/CEO of the creationist's organization, *Answers in Genesis* (AIG). According to the AIG website, Mr. Ham's ministry has the "purpose of upholding the authority of the Bible from the very first verse." Ham is the author of many books on Genesis, including "*The Lie: Evolution*," and he is widely sought after as a speaker.

The meeting Ham was addressing on this particular evening appeared to be in a relatively modest church building and the audience numbered three or four hundred people. The church was crowded, and the audience was extremely receptive to his talk. Although I found Ham's talk to be filled with misrepresentations and half truths regarding evolution, I was drawn to the congregation. There was a sense of oneness there, of security and identity. The atmosphere was comforting and supportive, a community of people who wanted and needed to belong to something that gave meaning to their lives. Singing and praying together with emotion they were reinforcing common beliefs, but more than that, they were reinforcing the bonds of their community. It was wonderful to watch.

Ham's talk was full of an "us versus them" approach. This approach plays well to people who feel victimized because in rising to the defense of their community and its values, their group identity is strengthened. It is unfortunate that the "enemy" in this case happened to be science, because it is a false enemy, one that in no way threatens religion. There are so many important issues regarding our social values and mores that could serve the same purpose for these good people. One does not have to think irrationally to maintain faith.

It is the need for group identity that leads to "group thinking," and the willingness to ignore any fact that is contrary to what we and our group believe. We like to think of ourselves as principled beings that sift the facts and make rational decisions, but the truth is that facts, or no facts, we believe what we want to believe. We listen most to people who affirm our beliefs, and shut out others, no matter how compelling their arguments. Ask a person today what news channel they regularly watch and you will have a pretty good idea of who they vote for, their position on abortion, their political party, and their slant on religion. It was group thinking that led the Dover Area School Board to try to force the teaching of intelligent design in biology classes even though none of the members had even the most rudimentary understanding of ID or of evolutionary theory.

We all need to be careful about defining ourselves and choosing which groups we wish to identify with. Most of all we need to look inside ourselves and constantly examine and challenge the choices we have made. Being chained for life to a dogmatic outlook greatly restricts whom we are and who we can become. The essence of life is challenge, growth and change.

CHAPTER 18

Life Happened !

The theory of creation science seems to be gaining ground among people who for one reason or another are opposed to the theory of evolution. It is an attempt to reconcile the biblical story of creation with modern biological concepts. The scientists who question the theory of evolution are primarily from the physical sciences. Chemists, physicists, and astronomers tend to have a mechanical, rather than biological, view of the universe. This way of looking at the universe is idealistic in the sense that it depicts nature at all levels of organization as conforming in design to some ideal. All water molecules, for example, are considered to be identical in structure and function whether they exist on earth, in a comet, or in a galaxy billions of light years away. A water molecule is a water molecule, is a water molecule. The same cannot be said about living things. In reality, every member of a sexually-reproducing species has a unique genetic makeup; one that is never repeated except among twins or in asexual reproduction.

Among humans, for example, each person is a unique genetic event, one that has never occurred before and in all likelihood, will never occur again. Try to imagine comprehending a universe in which every single water molecule was unique in structure. This would destroy our entire body of knowledge regarding chemical and physical processes.

Jesus died to take away your sins.
Not your mind
- Church Sign

Some people wield Scripture like a club. They see themselves as soldiers fighting to destroy the forces of evil, when in reality, they are simply people who cannot accept that others may not choose to think and act as they do. They cite scripture selectively that will back their viewpoint, often taking them out of context or misinterpreting their meaning. And they assume, without any apparent doubt, that they are justified in battering anyone who disagrees with them. There is no compromising with people of this frame of mind; they demand nothing less than absolute compliance with their views.

Design ?

In their book, *Rare Earth*, authors Peter Ward and Don Brownlee dispute the belief of many other scientists that given the enormous number of galaxies in the universe life must exist elsewhere in the universe. They point out that the factors which have enabled life to evolve on our planet are an almost inconceivable set of circumstances which are unlikely to occur elsewhere in the universe:

Our sun, while relatively small, has kept a steady temperature for the last 4 billion years.

Earth is the perfect distance from the sun for water to exist in a liquid form. Venus, which is much closer to the sun, has a temperature hot enough to melt lead, and Mars, which is much farther from the sun, is a frozen planet.

Our moon is the largest in the solar system, and it keeps the Earth's axis tilted at a steady 23 degrees. Without this stability, life would have difficulty adjusting to the constant perturbations as the planet wobbled around.

The Earth is geologically active and has a system of plate tectonics which is unique among the planets in our solar system. The movement of tectonic plates causes continents to move, mountains to raise, oceans to open and close, and regular volcanic activity. These relatively slow but constant geological changes have spurred the development of new life forms in a way that would be impossible on a geologically inactive planet like Mars.

Our solar system is in a perfect place in the galaxy. If it was nearer the center of the galaxy it would be exposed to extreme radiation and colliding and exploding stars. Life could not exist under those circumstances.

The planet Jupiter, with a volume 1300 times greater than the Earth's, produces a powerful gravitational field which acts like a giant vacuum cleaner sucking up asteroids and comets traveling through the solar system. If it were not for Jupiter, it is estimated that the Earth would be hit 10,000 times more often by these extremely dangerous objects. It has been estimated that the Comet Shoemaker-Levy 9 which collided with Jupiter in July, 1994, for example, released an amount of energy equal to detonating an atom bomb like the one dropped on Hiroshima every second for 13 years (In other words, more than 400 million bombs all detonated at once.) Had the comet struck the Earth instead of Jupiter, it would have destroyed most of the life on the planet and made it uninhabitable.

Ward and Brownlee refer to this special set of circumstances as *Rare Earth Factors*, and argue that despite the fact that there are an estimated 20 billion solar systems in our galaxy alone, the probability of these circumstances all coming together elsewhere in the universe are remote. They do not, however, argue for special creation.

While we must celebrate our uniqueness and do everything in our power to protect what may or may not be the only life forms in the universe, to suggest, as many creationists do, that this "uniqueness" could not have occurred without divine intervention is rejected by most scientists. Statisticians know that no matter how unlikely it is that a particular event will occur, given time it will occur, and having occurred, may never occur again.

Random events often happen in a way that seems planned or predetermined. Our minds constantly seek for patterns and often see them where they don't exist. The argument for intelligent design that is based upon the improbability of all of the circumstances falling into place which allows life to exist on earth is an example of flawed logic. One cannot interpolate backward from an event to show that it cannot happen by chance. For example, the odds of my picking the correct six numbers in a lottery using a 50 digit scheme are only one in 15,625,000,000. If I do pick the correct six numbers, can I then show that given the odds, my selection could not have been due to chance and that therefore some divine intervention has made me wealthy? Of course not. What I can do is to say *beforehand* that my ability to select the correct numbers is so unlikely that I can dismiss it as a likely possibility. Once it has happened, however, I cannot dismiss chance as a factor.

People who investigate accidents often comment that many accidents occur as a result of a series of unlikely events occurring all at once. Here is an example. My daughter Kate was preparing to leave on a three hour car trip. Just as she closed the door to her apartment and started down the hall she heard her phone ring. Hurrying back inside she was surprised to hear from a friend in Japan who she had not talked to in more than a year. The conversation lasted a little more than half an hour. Before getting on the interstate highway, she stopped for gas and went inside to pay and get a cup of coffee. While standing in line another customer bumped into her and the coffee spilled onto her blouse. She paid for her gas and went into the restroom to clean up as best she could. This took about ten minutes.

Once on the highway, she set the cruise control on 70mph and headed toward her destination. Meanwhile, about a hundred miles down the road, a large black dog bolted out the door of a farmhouse when the owner opened it to get the evening paper. Normally the farmer was careful not to let the dog out when opening the door, but this evening he was preoccupied by an argument with his wife. He yelled for the dog to stop and return to the house, but it fled into the night.

Meanwhile, back on the highway, my daughter suddenly saw the taillights of the truck ahead of her light up and she slowed. There had been an accident ahead, and traffic came to a standstill. Forty minutes later she was past the accident scene and on her way again. After another hour or so driving she was getting anxious to be at her destination and thinking about meeting her friends for dinner, when out of the corner of her eye she saw a fleeting glimpse of an animal bolting across the highway from the side of the road. Before she could swerve or brake the car struck the deer, and the animal was thrown into the median grass. Shaken, but unhurt, my daughter got the car stopped and pulled off the highway. A large black dog barked at her before running away as she got out of the car.

Is this a case of accident or design? What are the odds of all these individual events coming together to put Kate at that precise point on the highway at the exact split second in time when the deer shot across. If she had not received the phone call, she would have passed the accident scene more than thirty minutes before the deer approached. If the man had not spilled coffee on her, she would have missed the deer by ten minutes. If an accident had not occurred, she would have missed the deer by 40 minutes. If at any time during the trip she had accelerated or decelerated for even a few seconds, she would have missed the deer. If the farmer had not had an argument with his wife the dog would not have escaped the house and would not have chased the deer onto the highway.

There is no way to compute the probability of all these events happening in just the correct sequence and with the right timing to bring my daughter's car and the deer together in a 70mph collision, but I think you will see that the odds against her meeting a deer at that exact point on the highway again are so remote as to be considered impossible. And yet it did happen once. An argument for design based upon the improbability of the sequence of events leading up to the accident is logically flawed. So is the argument used to show design in the formation of life on Earth. By whatever series of improbable events, life did happen.

Does this mean that life resulted from a series of unlikely chance events that could never happen again. No, it means nothing of the kind. There may in fact be an organizing principle at work to create life when conditions are appropriate. The noted geneticist and evolutionary biologist Theodosius Dobzhansky once commented that he had searched his entire career for such an organizing principle, and Einstein tried vainly to develop a "theory of everything" (Unified Field Theory) that would explain the nature and behavior of all matter. The fact that these two great scientists were unsuccessful in their search does not mean that such natural principles are not at work. It only means that we have not yet reached a level of sophistication to recognize them.

The concept of intelligent design is perhaps the least intelligent concept since the belief that the Earth was flat and sailors who dared go too far out to sea would fall off into space. It is analogous to the sky is falling. It is throwing up your hands and whimpering that things are too complicated for our puny minds to comprehend, and therefore, some all powerful intelligent mind has to be behind it all. It is the same old story repeated countless times in the past to explain that which at the time was unexplainable. For example:

> Disease – God's way of punishing us for (use the appropriate one): (1) our sins, if we are not good people; (2) to test us, if we are disciples of God.
>
> Pregnancy – Children come from fertility gods. If we are in the god's favor we will be blessed with a son; otherwise we will be plagued with many daughters. In some parts of the world this ancient idea is still very much alive.
>
> Comets, meteors and "shooting stars" – Manifestations of the Creator. After all, they come from the Heavens, they must be omens from God.
>
> We are the beginning and end all of creation – The entire universe, all hundreds of billions of stars, planets, and other heavenly objects, were created for us, and our tiny planet is at the center of it all.

Children of God: Children of Earth

We are creators. We build buildings, bridges, vehicles, computers, and all the other vast array of technological tools upon which our survival depends. We do so very deliberately, with planning and testing of each product. We refine them through time, so that a computer in a few decades moves from several rooms in size to our pockets, or an airplane flight which lasts only 12 seconds and goes 120 feet leads to a landing on the moon in less than 70 years. We understand this process because it is the basis of our existence as a species, and it is as old as our species. We can imagine our ancestors sitting in a cave around a fire at night chipping away at a piece of stone with young children watching, children who would later try to make a better tool than their fathers taught them to make. It is not only ingrained in us by experience and teaching to think this way, it is genetically programmed into us.

And so, when we look at the complexity of the universe, or the complexity of DNA, cellular structure, or organism function, we naturally fall back on our heritage as creators to explain what we do not understand. We create things directly with our hands, and the thought that an automobile or refrigerator could come into existence through some indirect natural process is obviously ridiculous. We are creators, and because we are, it is easy to believe that an almighty creator is responsible for the world around us. The mindset that natural objects must also have been created directly by some supernatural being or beings is an ancient one, and it springs from our own innate ability to create. The scriptures tell us that we are made in the image of God. He is a creator and we are made in His image. The problem is that given this mentality, we will never understand natural phenomena.

If we accept creation science – God did it – then we have no need to continue the search for the origins of life on earth. This is an intellectual dead end as stultifying as accepting that all disease is caused by God to punish or test us. Beliefs like this may be emotionally comforting, but they will not keep our children from dying of disease, nor will they ever lead us to penetrate the Holy Grail of creation. As a biologist I have to admit that our concept of how life arose on Earth is too incomplete to be believable. Any biologist who says otherwise is simply being dogmatic. Accepting that we don't know is the first step toward the solution of any problem or enigma in science. Prior to Pasteur's discovery of bacteria, biologist had to admit that they simply did not understand the causes of disease, but they pressed forward in the belief that disease had natural, not supernatural causes, and in the end they broke through the ignorance of ages and made it possible for millions of people to survive who would otherwise have died while on their knees praying for help.

Scientists view the universe through a paradigm which focuses on natural causes. When we come up against something which we do not understand, we do not throw up our hands and say, God must be responsible. Scientifically (or in any other way), that is simply not a productive way of knowing about ourselves or the universe. Critics of science say this procedure is nothing more than a dogmatic belief which blinds scientists to the presence of a Creator. It is not a belief in the same sense that acceptance of God as Creator is a belief. It is a **method**, an approach to answering significant questions like what is the cause of disease or how did life originate on Earth. Roger Bacon once said that "To know what to ask is to already know half." Until the question "why do things fall to earth" was asked in a serious way, it could never be answered. But there is more to Bacon's statement. It implies that the question **can** be answered without resorting to supernatural answers, but asking the question is the beginning of the process, not the end. It is the start of a search for understanding, and if we are willing to accept the answer, because God made it that way, we will never really progress beyond our ancient ancestors who believed that spirits move the heavenly bodies and make crops grow.

Our knowledge of the dynamics of volcanism does not permit us to predict with any accuracy when eruptions will occur. People living in areas like San Francisco or Japan live with the knowledge that a big one is coming, but nobody knows just when. And yet just about everyone, from professional vocanologists to people on the street, believe that one day we will know enough to make predictions with accuracy. The concept of natural cause and effect is so ingrained in our society that most people trust science to come up with the answers that will save our lives from a volcano, disease, or an asteroid hitting the Earth. Why then in this one area, the creation of life and its evolution, do people who otherwise accept natural cause and effect, rebel? Why in this one area do they stubbornly cling to the superstitious and archaic beliefs of their ancient ancestors?

The answer lies partly with the failure of science to adequately provide a believable model for the spontaneous generation of life on Earth, and partly with our ancient reliance on a supernatural explanation for what we cannot understand otherwise. Before we mock this method of dealing with the unexplained, we should keep in mind that it has served us well, and is unquestionably one of the key factors in our species survival. What we need to realize, however, is it was not the Creator who sustained us but a belief in the creator. The supernatural is completely beyond our powers of comprehension, but our belief is personal and real to us. It is that which has always nurtured and sustained the human spirit.

We should also keep in mind that although organized religion has always been extremely conservative in accepting new ideas, it has always done so when confronted by science with concrete evidence. Yes, Galileo was persecuted, Darwin was ridiculed, and Pasteur was controversial, but in the end most organized religions accepted their revolutions when confronted with the facts.

In his acclaimed book *Ishmael*, Daniel Quinn points out that if evolution had not yet proceeded beyond the stage of jellyfish and their relatives, and if jellyfish were capable of reason, they would undoubtedly think that they were the end all of creation, and that nothing more complex would ever appear. They would also believe that everything that preceded them was leading to their appearance. He encapsulates this thought with the statement, "Then Came Jellyfish," as if jellyfish were the purpose and final goal of evolution, even though they were and are only an early step in the procession of life on Earth. That's how we think about ourselves. We find it nearly impossible to think beyond our own stint here on Earth, except in Biblical terms: the Earth will end and we will all be swept into Heaven. The thought that our species will disappear, as has 99% of all the species that have lived on the Earth, to be replaced with species not now in existence, is simply too much for our minds and egos to accept. And yet, that **is** exactly what will happen. We will one day go the way of the dinosaurs and the mammoths. The life of a species is like the life of an individual – it is born, develops and successfully lives for a period of time, and dies to be replaced by others, but life itself goes on in spite of all vicissitudes.

Change is a constant in our lives as it is in the life of the Earth. Nothing is more certain than that everything in the universe is constantly changing. This is often difficult to accept. It seems to be part of our nature to want to hold onto the moment, to find a secure and happy place and live in it forever. Our concept of Heaven is based upon this need. Yes, our bodies are slowly failing as we age; loved ones who have been a part of our core world for as long as we have lived pass away and leave a void that can never be filled; technology changes our lives at a pace that often leaves us breathless and confused; problems mount upon problems until we feel overwhelmed; but, in the end we will find that good and perfect life in Heaven, a life that never changes in its perfection. It is this which often sustains us, but it is also this which causes us to reject changes in our religious and scientific beliefs. We often prefer absolutes that do not exist: the Bible is the word of God as given to the ancients and it is eternal and infallible; the Earth was created in six days and is only a few thousand years old; God created all species of plants and animals and they have remained unchanged since the creation; the Earth is the center of the universe; all of creation was made by God for his glory and for the benefit of mankind. Thinking like this does not lend itself to a concept of a changing earth or universe, and those who think in these terms hold to their beliefs very stubbornly.

James R. Curry

Project Steve

As a parody to the ongoing efforts of evolution deniers to create lists of "scientists" who are opposed to evolution, the National Center for Science Education issued the following statement and asked scientists to sign it:

Evolution is a vital, well-supported, unifying principle of the biological sciences, and the scientific evidence is overwhelmingly in favor of the idea that all living things share a common ancestry. Although there are legitimate debates about the patterns and processes of evolution, there is no serious scientific doubt that evolution occurred or that natural selection is a major mechanism in its occurrence. It is scientifically inappropriate and pedagogically irresponsible for creationist pseudoscience, including but not limited to "intelligent design," to be introduced into the science curricula of our nation's public schools.
- National Center for Science Education Project Steve

At the time of this writing, 983 scientists have endorsed the statement. Called Project Steve in honor of Stephen Jay Gould, only scientists named Steve are eligible to sign. "Steves" make up about 1% of all scientists.

There is always a conflict between those who feel their values are eternal and those who feel they are relative
- Richard Kostelanetz

Science and religion (or other beliefs in the supernatural) are not mutually exclusive human endeavors. On the contrary, most scientists believe that they are complementary ways of understanding the universe and our place in it. The problem lies in two very different paradigms of the world: the scientific paradigm, which holds that no concept or principle is absolute; and the paradigm of absolutism held by so many Christians and peoples of other faiths. People who hold that the word of God as written in the Bible or other religious texts is unalterably true for all time have a rigid mind set that makes them feel threatened by and suspicious of scientists who are constantly examining and questioning their own principles and beliefs. Holding to absolute truths themselves, they tend to dismiss the credibility of any concept whose validity is in question or for which there are no concrete solutions or answers. It is impossible for them to accept the fluid, constantly changing world of ideas that scientists revel in. To their minds, beliefs which must be challenged and tested daily are no beliefs at all. Salmond Rushdie, who went into hiding after his life was threatened by Iran's Ayatollah Komeini following the publication of his book, *Satanic Verses*, contrasts these two different world views in the following quote:

> What is a novelist under terrorist attack worth?
> Despair whispers in my ear, "Not a lot." But I
> Refuse to give in to despair.
> Our lives teach us who and what we are. I
> have learned the hard way that when you permit
> anyone else's description of reality to supplant
> your own – and such descriptions have been
> raining down on me – then you might as well be
> dead.
> Obviously, a rigid, blinkered, absolutist world
> view is the easiest to keep hold of, whereas the
> fluid, uncertain, metamorphic picture I've always
> carried about is rather more vulnerable. Yet I
> must cling with all my might …to my own soul;
> must hold on to its mischievous, iconoclastic,
> out-of-step clown instincts, no matter how great
> the storm. And if that plunges me into contradiction
> and paradox, so be it; I've lived in that messy ocean
> all my life, I've fished in it for my art. This turbulent
> sea was the sea outside my bedroom window in
> Bombay. It is the sea by which I was born, and which
> I carry within me wherever I go.
> "Free speech is a non-starter," says one of my
> Islamic extremist opponents. No sir, I reply, it is not.
> Free speech is the whole thing, the whole ball game.
> Free speech is life itself.
> What is my single life worth?
> Ladies and gentlemen, you must decide what you
> think a writer is worth, what value you place on a maker
> of stories, and an arguer with the world.

Quoted from N. Strossen, ACADEME: Jan/Feb, 1993

Environmental Values

Our values regarding the natural world changed radically with the introduction of agriculture and the domestication of animals approximately 12,000 years ago. As hunter-gatherers we worshiped nature in all its forms and sought to conform to the natural systems which sustained us. We competed for survival, but did not make war on our competitors. As agriculturalists we began to look upon nature as capital, something which could be controlled, owned, bought and sold, an enemy to be conquered and if necessary, destroyed. Our attitude toward predators is an example. To hunter-gatherers the eagle, wolf, tiger, and other predators were animals to be respected, emulated, and in many cases worshiped for their prowess and courage. To the agriculturalist they were threats which had to be attacked and completely eradicated. The same was true of plant-eating insects, rodents, poisonous animals of all kind, plants which competed with crops (weeds), "creepy-crawlers" like spiders, and a host of other plants and animals.

The environmental dilemma we face is on a scale unlike any threat ever faced by our species. When hunter-gatherers degraded their local environment they simply moved on to more productive fields and allowed their former haunts to recover. That option is not available to us: we cannot move to new, fertile planets. We face environmental challenges and threats today never dreamed of by our ancient ancestors. The following two chapters explore those challenges and discusses the role science and religion must play if we are to successfully deal with them. Nothing less than our continued existence as a species is at stake.

CHAPTER 19

Rapa Nui

A formidable challenge lies ahead: the rapidly growing demands of human society vs. the rapidly diminishing resources of the natural world.

- Harry Hollis

On a brilliant sun-lit day some 1600 years ago a double-hulled canoe filled with voyagers moved easily over the calm waters of the blue Pacific. The oarsmen moved in unison, chanting as they worked. Suddenly one of them cried out and pointed to a tiny tip of green projecting above the blue water. They all paused for a moment and cheered excitedly, then bent to their work with renewed vigor until the boat fairly leaped over the waves. Several hours later they stepped onto the beach of one of the most isolated islands in the world and stretched their stiff legs. The warm ocean breeze felt cool on their wet backs as they stared in amazement at the emerald green island rising above them.

These were the first colonists to Rapa Nui, called Easter Island by Europeans, and they had found an uninhabited paradise with fertile soil, abundant food, and plentiful building materials, surrounded by a bounteous ocean. Over time, the colonists prospered and grew in numbers. They developed a system of agriculture which coupled with harvesting of the ocean provided them with a plentiful source of food. As their wealth grew, their classless society became tiered, with an elite ruling class and classes of artisans, farmers, and seafarers. With abundant food came leisure time and they began to cut enormous stone figures from the native volcanic rock, and move them to places overlooking the ocean. These figures grew ever larger as time passed until figures more than 30 feet high and weighing nearly 300 tons were being carved and moved up to eight miles from the quarries. Eventually, more than 200 statues stood on stone platforms facing the sea, while hundreds more were abandoned at the quarries and in the fields.

In 1774 when the Dutch explorer Jacob Roggeveen stopped at Rapa Nui Island on Easter day, the islanders had no contact of any kind with the outside world and were unaware that other people existed apart from themselves. The island that Roggeveen found was far different than the original one the colonists had discovered centuries earlier. The thick forests were completely gone, replaced by sparse grasses which seemed more dead than alive, and the people no longer had enough wood to build fires or construct boats. The palm trees which had once dominated the forests were extinct, as were all the species of land birds that had originally existed there.

Moai on Easter Island

When the first settlers arrived on the island large numbers of seabirds nested, including terns, tropic birds, boobies, frigate birds, albatross, shearwaters, storm petrels, and fulmars, and the colonists used them as food. According to Jared Diamond in Discover Magazine (August 1995), the island was "the richest seabird breeding site in Polynesia and probably in the whole Pacific." By the time of Roggeveen's visit, more than half of the seabird colonies had been destroyed according to Diamond. The number of islanders at this time was estimated at fewer than 2000, most of who were living in poverty and near starvation. Recent estimates by archeologists put the number of people at the height of Easter Island's prosperity during the 9th century AD as high as 20,000.

Once the forests were gone the islanders were no longer able to construct suitable boats for fishing and catching porpoises, and they were more or less relegated to what they could catch along the coasts. Unfortunately, the coastal waters became depleted from overuse. They then relied heavily on chickens and rats for meat, and eventually turned to cannibalism. The plentiful remains of human bones in garbage heaps are testament to the desperate condition to which the once thriving society had fallen as a result of the collapse of the island's resources.

Under the stress of starvation their social system broke down. Central authority was no longer able to exert control and the people splintered into small clans that raided and warred against each other for resources that would sustain life. By the 16th century people were hiding in fortified caves to protect themselves from the continuous warfare and raiding which went on between clans. Lacking wood to build boats, they were trapped on an island only 11 miles long and 7 miles wide. Perhaps out of frustration and anger they knocked over each others statues and abandoned work on hundreds of others which were under construction or in the process of being moved to the coast. In so doing, they abandoned the cult of ancestor worship which the statues represented. Small hand-held wood statues have been found from this period which depict starving individuals and contrast sharply with the huge, obese stone statues of earlier years.

The desperate situation of the late 15th and early 16th century was set in motion centuries earlier by the inability of the islanders to appreciate that the island's resources were limited and had to be used in a sustainable way if their society was to flourish indefinitely. They discovered a land untouched by human hands, pristine, fertile, and full of exotic plants and animals, a Garden of Eden. They came into this tiny world full of hope, and for awhile their hopes were fulfilled – they prospered and multiplied. But then they began to abuse the garden with no sense that they were sowing the seeds of their own downfall. This is a pattern that has been repeated many times in human history. It happened to the first civilizations in the Middle East, to the ancient Egyptians, to the Maya and other Central American indigenous civilizations, and it would have happened to the countries of Europe if a whole new world rich in natural resources had not been discovered and colonized in the Americas.

The Polynesians who colonized Rapa Nui lived close to nature and were sensitive to even the most subtle changes in their environment. How is it that they could not anticipate the eventual destruction of their island world? The same question can be asked about all of the other cultures which have fallen from prosperity to decline because of overpopulation, overconsumption, and environmental degradation.

I believe the answer is that agriculture thrust our ancestors into a completely different set of environmental circumstances than those they evolved to deal with. As hunter-gatherers people lived in small tribal groups and existed from day to day on what they could find to eat. This lifestyle required that they look no farther to the future than what they would eat today or tomorrow. There was no need to think and plan years or decades ahead. Coupled with this real time mentality was the imperative to amass and consume as much food and other resources as possible when they were available because what was abundant today could be impossible to find tomorrow.

The development of agriculture completely changed the way humans relate to the environment. Agriculture is an extreme form of environmental manipulation, and it empowered the human race as it had never been empowered before by providing a much larger and more reliable source of food, much of which could be stored for future use. Never before had people had such control over their food source and their own lives. Agriculture tied people to the land and gave them a sense of ownership over nature which had never existed before. They were no longer a part of the land; the land belonged to them. This concept of ownership eventually spread to all of nature, until people began to think of the natural world only in terms of ownership: our land, our air, our oceans, our forests, our rivers and lakes, our mountains, our animals, our plants, our Earth.

The abundance of food provided by agriculture allowed population sizes to increase from a few million world-wide to tens of millions and then billions. As wealth increased with food production social classes formed and governmental structures developed to regulate resource use and trade. The ruling classes and elite soon began to do their best to out-compete each other with lavish displays of wealth and power. With abundant food large numbers of people were freed from farming, and classes of artisans, builders, shamans, and traders, sprung up. Resource use became ever more extravagant in the building of temples, houses, governmental buildings, (or in the case of Rapa Nui, statues), and in the use of personal adornments like clothing and jewelry. These two factors - over-consumption of limited or nonrenewable resources, and unchecked population growth - eventually lead to collapse of the economic systems, warfare with neighboring peoples for resources, and eventual collapse of society. This is the history of the human family over the past 12,000 years, all brought about by behaviors which were essential for survival in a hunting-gathering world but which were maladaptive in an agricultural one.

Stone-age people faced challenges every day which tested their ability to survive, and they were obviously very good at meeting those challenges or we would not be here today. They were not required to look far into the future and consider the consequence of their actions or inactions. Genetically, they, and we, are not programmed to deal with future problems. When faced with immediate challenges – destructive storms, earthquakes, military attacks, September 11, an automobile accident, we react immediately and effectively. When faced with possible future challenges, we react, if at all, with indecision and delay. There were plenty of warnings about the possibility of terrorists using aircraft to attack buildings in the U.S., but we were ineffective in dealing with the threat. For years people in the know warned that a category 5 hurricane would flood the city of New Orleans, but nothing was done to strengthen the levees. We are now being warned of the effects of global warming, but our mentality is not capable of becoming alarmed at events which may happen decades or centuries into the future. We deal with the here and now, and push thoughts of the future into the dark recesses of our mind. What concerns us is the need to attend to the multitude of small problems which each day brings.

How many people lie awake at night worrying about south Florida disappearing under the waves of the ocean as a result of global warming? No, we lie awake at night worrying about paying our bills, keeping up with the Joneses, out-competing our colleagues for advancement at work, or problems with the kids. Events decades or centuries away are of little consequence to our Stone-age brains.

There are many similarities between what happened on Rapa Nui and what is happening to the world at large today. We too are trapped on an island from which we cannot escape. Our island, like theirs, has limited resources, and like them we are beginning to war over resources that are in short supply. Ours is an island in space. We cannot leave it and we cannot get food or other resources from outside. And like the people of Rapa Nui, we may have already sown the seeds of our own downfall.

The people of Rapa Nui destroyed most of the plant and animal life on their island and in its coastal waters, a source of food which they needed to survive; we are destroying ours at the rate of a hundred or so species a day. In 2005, more than 2000 of the world's leading environmental scientists from nearly 100 nations completed the *Millennium Ecosystem Assessment*. They concluded that the current global extinction rate is 100 to 1000 times greater than normal and projected that the rate will increase tenfold or more in coming decades.

The Rapa Nui people eliminated all of the forest on their island; we too are eliminating our forests at an unsustainable rate. In 1990 the Food and Agricultural Organizations of the United Nations estimated that the total forested land area was approximately 3400 million hectares. The loss of forest area during 1980-90 was estimated at 163 million hectares, of which 154 million hectares was in the tropics. The sky-rocketing birth rates in tropical countries and consequent pressure to destroy forests to provide land for small farms, and the rapid industrial development of countries like China, ensure that this trend will not only continue, but increase. The exporting of wood products from China in the form of inexpensive furniture and other products has increased 900% in the last eight years. Most of this wood comes from tropical countries where much of it is harvested illegally. As a result of population growth and the destruction of forests in developing areas of the world, the amount of forest area per world capita fell from 3 acres in the 1960's to 1.5 in the 1990's, and it is projected to fall below ½ an acre by 2020.

The Rapa Nui people over harvested the coastal seas and were reduced to feeding on seafood that was less nourishing and more difficult to find; we too have depleted all of the most productive fishing grounds in our oceans and are reduced to catching and feeding on less desirable seafood. Within the past 50 years our fishing practices have reduced the oceans population of edible fish by 90%. Ocean productivity reached a peak in the 1970's with the harvesting of some 120 million tons per year. By 2006, despite new technology such as the use of sonar, satellites, and nets as long as 18 miles, this declined to less than 92 million tons and countries that used to be exporters have become net importers. The efficiency of today's fishing technology and practices insure that this downward trend will continue even while the world demand for seafood continues to soar.

The Rapa Nui allowed their population to grow larger than their island could sustain. The result was a crash from 20,000 well fed people to less than 2,000 nearly starved individuals. The total world population today is more than 6.5 billion people and it is still growing. A billion is a thousand million. To put this into perspective, if a person were to count once each second without ever taking a break for sleep or anything else, it would take more than 30 years to reach a billion. As mentioned earlier, for most of our history as a species the total world population was around 5 million. We did not reach the one billion mark until after 1800. In other words, it took all of human history until the 1800's for the population to grow to a billion. It took less than 130 years to add the second billion, which was reached in 1930. The third billion was reached in 30 years (1960); the fourth took only 15 years (1975); the fifth 12 years (1987) and the sixth 12 years (1999). Clearly we cannot continue to add a billion people to the earth every twelve years.

Global Warming

During the last century the number of people on Earth more than tripled, from less than 2 billion to more than 6 billion! If that same rate of increase occurs during the current century, the world will enter the 22 century with nearly 20 billion people. Does anyone really believe that the earth's top soil, oceans, fresh water, forests, mineral and energy deposits, all of which are already stressed, can support 20 billion people?

When the last trees on Rapa Nui were gone, the people had to turn to grasses and sedges to cook their food and heat their houses. Today 40% of our global energy use is from crude oil. Some oil industry analysts calculate that we have extracted half of the world's oil reserves. In 1956, Shell Oil geologist M. King Hubbert calculated that U.S. oil production would peak in 1970. Although he was ridiculed for his prediction, it came true and production has declined every year since. In 1974 he predicted that global oil production would peak in 1995. It did not, but many scientists have improved upon his technique and believe that he was not far off in his estimate. At current levels of production and use most analysts estimate the world's oil supplies will last about 40 more years.

Despite the denials of business leaders and economists who wish to maintain an ever expanding economy and a life style based upon overconsumption, the planet's atmosphere *is* warming, most likely as a result of human activity. Carbon dioxide and other "greenhouse" gases reached record levels in the atmosphere in 2004 according to the World Meteorological Organization. These gases are primarily the result of the burning of fossil fuels like gasoline, diesel fuel, natural gas, propane, coal and wood. When these carbon compounds are burned by chemically breaking them down with oxygen, carbon dioxide is one of the products. Water vapor, nitrous oxide, methane, and a few other gases are also considered to be greenhouse gases.

The name greenhouse gas comes from the ability of these gases to trap and hold heat in the atmosphere. Light can pass through the glass of a greenhouse or closed automobile, but when it strikes the ground or objects inside it is converted to heat, which is then transferred to the air. Since the air cannot escape, the temperature inside the greenhouse or car rises. The same thing happens with greenhouse gases like carbon dioxide. They allow solar radiation to pass through the atmosphere which then strikes the surface of the earth and is converted to heat. This heat is transferred to the air and is trapped there by the greenhouse gases. Anyone who has ever sweltered through a muggy July night knows that the temperature seems not to fall at all when the sun goes down. This is because the high water content (humidity) of the air traps the heat near the ground. In the dry air of a desert, on the other hand, temperatures begin to fall as soon as the sun goes down and by morning it can be quite cold. This is how greenhouse gases of all types work.

Carbon dioxide is the primary greenhouse gas. The planet Venus is a good example of what can happen when carbon dioxide levels in the atmosphere are high. The temperature of Venus's atmosphere reaches almost a thousand degrees and would be higher if it were not for a thick cloud cover which reflects much of the solar radiation back into space. NASA has reported that the Earth's surface temperatures during 2005 were the highest since recordings were begun in the late 1800's. Scientists who have bored deep into glacial ice and measured gases trapped there for centuries found that the atmosphere today contains 35% more carbon dioxide than it did in 1750.

A 2006 study by scientists from the University of Arizona and the National Center for Atmospheric Research reported in the Journal *Science* that sea levels are rising faster than previously thought and could rise as much as 13-20 feet by 2100. A rise in sea level of even a few inches could have a devastating effect on coastal towns, cities, and ecosystems world wide as these are some of the most heavily populated areas in the world. In the United States, for example, 53% of all residents live in coastal areas.

Island nations will also be affected by rising sea levels. The Pacific Island nation of Tuvalu is losing 3.5 inches of elevation per decade and many of its 11,000 citizens are fleeing to New Zealand, where they are accepted as environmental refugees.

Sea ice in Hudson's Bay is vanishing two weeks earlier than it did as recently as the 1970's. A comprehensive international study in 2004 predicted that one half of the Arctic's summer ice would vanish by the end of the current century, and fast moving glaciers in Greenland now spill twice as much ice into the Atlantic Ocean each year as they did in 1996. As ocean temperatures rise, the number and intensity of hurricanes and cyclones will also increase because these storms gather their energy from heat stored in the seas.

Coral reefs everywhere are dying from the effects of global warming and human interference. Living coral animals form a thin skin over the dead rock-like remains of coral which lived and died before them, sometimes building extensive reefs. Australia's Great Barrier Reef at 1300 miles long is the only living thing that can be seen from space. A recent study by Queensland University's Centre for Marine Studies predicted that as a result of global warming the Great Barrier Reef will lose most of its coral cover by 2050, and in a worst case scenario, the reef will face destruction by 2100 regardless of what actions are taken to prevent it. The study also concluded that it is very unlikely that coral reefs could be reestablished over the following 200-500 years. Reefs are among the world's most productive and diverse ecosystems, and their loss would be incalculable. According to the U.S. National Oceanic and Atmospheric Administration (NOAA) coral reefs provide food, income, jobs, and other important services to billions of people and the social, economic, and environmental consequences of reef destruction is immense.

Deforestation also adds to the buildup of carbon dioxide in the atmosphere. Green plants remove carbon dioxide from the air and combine it chemically with water to form simple sugars. These are then converted into all of the other materials of the plant. When trees are cut and burned all of the carbon compounds in their tissues are released into the air as carbon dioxide. The destruction of forests is doubly harmful in that trees that would remove carbon from the air are gone, and the carbon that was tied up in their tissues is released into the air.

Desertification

The nation that destroys its soil destroys itself.
Franklin D. Roosevelt

In 1935, Paul B. Sears published a book called *Deserts on the March* which first brought the problem of desertification to the world's attention. Desertification is land degradation in arid areas of the world caused by human abuse and climate change. According to former United Nations Secretary-General Kofi Anan, desertification is "One of the world's most alarming processes of environmental degradation." The U.N. Convention to Combat Desertification (UNCCD) estimates that "one third of the earth's surface and over a billion people are affected. Moreover, (desertification) has potentially devastating consequences in terms of social and economic costs."

Desertification is occurring today in more than 100 countries. Nearly 70% of the world's useful dry-land for agriculture has suffered soil erosion and degradation. When productive land is turned into desert as a result of over-cultivation, deforestation, or overgrazing, people are forced to leave their farms for jobs in the city. Crowded into shanty towns without proper services or facilities, they live in abject poverty amidst filth and violence. Desertification is devouring more than 20,000 square miles of land worldwide every year, which is a contributing factor to the deaths of 30,000 people who starve to death every day.

Desertification is not the only cause for the loss of arable land. In the United States from 1982 to 1997, 25 million acres (39,000 square miles) of productive farmland was turned into housing subdivisions, malls, workplaces, parking lots, resorts, and other non-farming uses. According to the American

Farmland Trust, we lost farm and ranch land 51% faster in the 1990's than in the 1980's and we are losing the most fertile and productive land the fastest. This necessitates putting marginal land into production, which increases the need for water and farm chemicals to make them productive. Two acres of farmland are lost in the U.S. with every passing minute. If this rate continues to the end of the 21st century the acreage of land growing crops for each U.S. resident will decline from 1.4 acres/person to 0.46 acre in 2100. This in turn will put even more pressure on the remaining acreage and accelerate its decline.

"Soil erosion is second only to population growth as the biggest environmental problem the world faces," according to David Pimentel, professor of ecology at Cornell. "Yet the problem, which is growing ever more critical, is being ignored …" If the process is allowed to continue what will be left for people a thousand years from now? Does anyone think that far ahead or worry about what our descendents will do for food?

Living in the Midwest amongst some of the world's most fertile soil, I have the opportunity to watch as each year more and more farmland disappears to development. Sometimes when I am passing a field that is being worked by a farmer I see the bare soil being blown across the road or washed away by rainfall and wonder how such land will still be productive a hundred, five hundred, or a thousand years from now, and question whether anyone really cares. Destructive farming practices will have to change if future generations are to enjoy the lifestyle we enjoy today.

Next to water, topsoil is the most precious and non-replaceable resource on the planet. When the people of Rapa Nui destroyed the productivity of the soil, their population plummeted from 20,000 to less than 2,000, most of who were near starvation. Like the Rapa Nui, we live on an island which has limited arable land, and like them, agriculture is the base of our economic system. When the base collapses, economic ruin, starvation, and the breakdown of social order follow. Farmland in the U.S. and elsewhere in the world should be protected from all forms of development which will take it out of production. To do anything less is to threaten the survival of future generations.

The growth of cities into rural areas is a major factor in the loss of farm land worldwide. Between 1987 and 1992, for example, China lost nearly two and a half million acres of farmland each year to urbanization and the consequent expansion of roads and support industries. In the US estimates are that urban sprawl consumes nearly one million acres each year. UN estimates project that by 2015 the world's urban population will be nearly 60% (up from 45% in 1995). Urbanization will undoubtedly continue to swallow up farm land well into the future.

Water

The fires of life burn only in water

There was a time not so long ago when no one had ever seen what the Earth looks like from space. Artist's depictions looked more like a schoolroom globe than a sparkling blue gem set in a sea of blackness with white wisps swirling about it like the dew-covered gossamer scattered across a meadow on a summer's morning. The island nature of our planet becomes clear when viewed from afar, for outside the thin layer of atmosphere lies everything that is antithetical to life: extreme cold, killing radiation, and the complete absence of any life support systems. The Earth is blue because it is a water planet, and because the heat exchange between the planet and space is balanced to maintain a climate where water can exist in its liquid form. If this were not so life could not have evolved or prospered here.

> *The wars of the twenty-first century will be fought over water.*
> Ismail Serageldin - World Water Commission Chairman

Water is the lifeline of human existence, and consequently the most precious resource on the planet. Historically civilizations have always flourished where fresh water was in plentiful supply for use in drinking, cooking, bathing, and irrigation. Following the industrial revolution, rivers, lakes, and streams took on a new importance, for large amounts of water are required in most industrial processes or to carry away the wastes of production. As cities and towns grew in number and size these same waterways became conduits to carry away ever increasing amounts of sewage and other wastes.

Only 3% of all the water on Earth is fresh water. Approximately 79% of this is tied up in ice caps and glaciers, and another 20% is underground. The remaining one percent is in rivers, streams, and lakes. In other words, only .03% of the Earth's water is readily available as surface water, and this is very unevenly distributed. Much of it is also heavily polluted. Forty percent of all lakes and streams in the United States are too polluted to use for fishing or swimming. Heavy metals like mercury and lead, PCB's, agricultural chemicals, disease organisms, heat, oil, silt, mine tailings, household sewage, medical wastes – these and many other materials have turned many of our streams and lakes into toxic brews, and many of them will remain so for decades or centuries.

According to the Environmental Protection Agency, one third of U.S.

lakes and a quarter of its rivers are contaminated with high enough levels of mercury to cause health problems for children and pregnant women. Eighteen states have issued warnings about eating fish caught from all lakes and rivers. The mercury comes primarily from power plants that burn coal. Mercury-tainted fish can severely damage the nervous system of children and unborn fetuses.

Our lives would be impossible without a plentiful supply of fresh water. We build our cities near water and our economies would collapse without it. Water is our life and our culture, but because we continue to treat it as just another consumer product, we are creating a world water crisis that has the potential to depress the world's economies and bring about violent conflict.

In his message on the World Day for Water in 2006, former Secretary General of the U.N. Kofi Annan had the following to say:

> *An estimated 1.1 billion people lack access to safe drinking water, 2.5 billion people have no access to proper sanitation, and more than 5 million people die each year from water-related diseases — 10 times the number killed in wars, on average, each year. All too often, water is treated as an infinite free good. Yet even where supplies are sufficient or plentiful, they are increasingly at risk from pollution and rising demand. By 2025, two thirds of the world's population is likely to live in countries with moderate or severe water shortages. Fierce national competition over water resources has prompted fears that water issues contain the seeds of violent conflict.*

China is a good example of how industrial development and population growth have led to increased demands for water. The Yellow River, once the cradle of Chinese society, is now reduced in many places to a sluggish, toxic stream which dries up before reaching the ocean for much of the year. As much as 90% of the water is used for irrigation by farmers along the river. The decline in the supply of water for China's farmers is due to the rapid growth of cities and industry and pose a threat to the world's food supply, according to the World Watch Institute. The drying up of rivers and depletion of underground water is threatening China's ability to produce enough food for its people. The World Watch Institute has warned that if China has to resort to importing food, world grain prices will soar, causing social and political instability in Third World countries.

Much of the world's water is underground (95% in the U.S.), but underground water often exists as a non replenishing pool which is not replaced when it is pumped out of the ground. The ground above these underground pools (aquifers) sometimes subsides as the water is pumped out. According to the BBC News web site, farmers in the Texas High Plains are pumping groundwater faster than it is being replenished. As a result, North America's largest aquifer, the Ogallala, which stretches from Texas to South Dakota, has been depleted by a volume equal to the annual flow of 18 Colorado Rivers. Mexico City is sinking by nearly an inch and a quarter per year due to the massive reduction of water in the aquifer underneath the city, and with the population of the city increasing daily this will accelerate.

The most sacred Hindu river, the Ganges, is so depleted according to reports by the BBC News that the wetlands and mangrove forests of Bangladesh are seriously threatened. The Ganges also contains unacceptably high levels of arsenic and other pollutants.

A U.N. report predicts that access to water may be the single biggest cause of war in Africa in the next 25 years. Dams being built by Turkey along the Tigris and Euphrates Rivers have caused Syria and Iraq to accuse that country of depriving them of badly needed water.

The Aral Sea in Central Asia lies in the center of a large, flat, desert basin. It was the world's fourth largest lake in 1960, but has since decreased by 75% due to human induced environmental changes. The shoreline has receded up to 72 miles from its former shore. The amount of water lost is the equivalent of draining both Lake Erie and Lake Ontario.

In the Middle East, water is more important than oil. Competition for water from the River Jordan was a major factor in the 1967 war and continues to aggravate regional tensions. According to Friends of the Earth Middle East, 90% of the water from the river is being depleted by Israel, Jordan, and Syria, and what remains is heavily polluted. Friends of the Earth estimate that the river could run completely dry within two years. The pollution from the Jordan River flows into the Dead Seas, which has shrunk by 30% in the last few decades.

In East Africa competition for dwindling water has fueled increased violence among rival nomadic tribes. The lack of rain is threatening nearly 8 million people with starvation. In some regions more than 80% of the cattle have died, and people are forced to travel long distances in search of food and water. In February, 2006, nearly 40 people were killed after a Kenyan tribe drove its cattle into an area claimed by Ethiopians. According to a piece in the Los Angeles Times, clashes like these have caused alarm in a region awash with grievances and guns.

The population of the world continues to grow; the Third World Countries continue their efforts to develop; and the wealthy countries of the world continue to increase their consumption of water for recreation, fountains, beautiful lawns, reflecting pools, retirement communities in the desert, car washers, and a host of other nonessential uses. We have been warned by the U.N and other experts that the world is facing a water crisis which could lead to growing violence, economic collapse, and starvation on a massive scale.

In the U.S. the governors of eight Midwest states signed a compact in 2005 which would prohibit piping or shipping water from the Great Lakes to other states, regions, or other parts of the world. There is strong support in Congress to pass legislation supporting the compact. The governors were moved to act after a Canadian firm got a permit from Ontario to ship tankers of water from Lake Superior to Asia.

Rush to Ruin

We are following the same path as the people of Rapa Nui, and even though we are in a much better position to recognize what we are doing and have the means to make corrections, our efforts are pitifully weak and too slow to really stop our progress toward a breakdown in our ecological and social systems. As resources become limiting, starvation and poverty will increase, resources will be fought over, governments will fall, and our Garden of Eden could become a living hell. None of this need happen, but given the apparent lack of movement toward a truly sustainable use of the Earth's resources, it would appear that the lessons of the past are falling on deaf ears. Scientists in every discipline are warning of the tragedy that will result if we continue to ignore the signs of resource depletion and environmental degradation, but we in the wealthy and powerful countries are too preoccupied with maintaining and displaying our wealth and power to take effective action.

I have no doubt that if someone had warned the Rapa Nui about their worsening situation and the tragedy that lay in wait for them two or three centuries before their collapse they would have reacted as we are and continued to carve out ever larger statues (Moai) and destroy their forests, soils, animals, and marine resources. Events that may happen centuries into the future do not ring the human crisis alarm. Today's Moai are the wasteful vehicles we choose to drive, extravagantly large houses, increasingly large buildings, opulent life style, and wasteful use of resources; all designed to flaunt our wealth while despoiling the environment and depleting vital resources. The relationship between the attainment of material security and human satisfaction is very weak, if it exists at all. More is never enough.

Not only is more never enough, but there seems to be no relationship between "more" and human happiness. Regular surveys by the National Opinion Research Center of the University of Chicago show that the percentage of Americans reporting that they are "very happy" is the same now as in 1957, despite a near doubling in personal income and consumption expenditures. The peoples of the world have consumed more goods and services since 1950 than all previous generations put together, but report that they are not any happier.

The peoples of the developed countries are consuming resources at an ever increasing rate, and the peoples of the underdeveloped countries are attempting to develop at the same time that their populations continue to increase. These two factors, population growth and the growth of consumerism, have the potential to cause a breakdown in the world's cultures if left unchecked. Already there are signs of this happening as wars over resources rage and tens of millions of people from the underdeveloped areas of the world flood into developed areas and create social conflict. I expect this conflict to intensify as resources become more scarce, and as a result, more unevenly distributed. The elite class of people on Rama Nui insisted on continuing their lifestyle of greed and waste until it was too late. I see the same thing happening on a global scale today. It is human nature – the nature of a Stone-age creature that used gluttony to hoard scarce resources when they were available against those days when they were not, and to display his power by flaunting what he had accumulated. It served him well; it may destroy us.

CHAPTER 20

Then Came Christianity

I was not born to be of this world.

- E.O. Wilson

"The most dangerous of devotions…is the one endemic to Christianity. *I was not born to be of this world.* With a second life waiting…the natural environment can be used up." This quote from Wilson's book *Consilience: The unity of Knowledge*, attacks the Christian concept of the Earth as a temporary residence, a place to be exploited while we wait for the return of the Savior. Environmentalists and ecologists think of the Earth in terms of cycles: plants and animals die; the life-giving chemicals of their body decay and are returned to the soil, where they are picked up by other life forms and once again become part of living matter. This cycle has gone on endlessly from the beginning of life on Earth and will continue endlessly so long as life exists. All of the natural processes, physical as well as living, have their own impetus and will continue until our sun is used up and dies, but that will not be the end of all things natural, it will only be a new beginning.

For Christ died for our sins and rising dies no more.

Christians, on the other hand, see time as linear, non-repetitive, and preordained by God. The Earth has a beginning and an end: in between is the unfolding of a Biblical drama: Jesus Christ is born, dies, and is resurrected. He came to save us from our sins, and He will return, at which time the Earth will be consumed by fire. The end. It makes perfect sense not to be concerned about the Earth and its resources if you buy into this concept, especially if you believe, as many Christians do, that the return of Christ will happen soon. We are all going to a better place where the cares and trials of life give way to perfect harmony and eternal peace, so why be concerned about a physical world that is corrupt and of a lower order of existence? The world is but a staging area of temporary utility and has no long term importance in Christian dogma.

That all human beings are created in the image of God is central to Judeo-Christian belief. This distinguishes humans from all other living creatures. When Christianity destroyed pagan animism, it also destroyed our reverence for the Earth as our mother. We no longer think of the soil we walk upon as being sacred, even though it nourishes us and provides us with the very means of existence. It has become simply a commodity to be bought and sold, and the earth itself is there to be exploited, even it means despoiling it. Langdon Gilkey in his book, *Nature, Reality, and the Sacred*, puts forth the belief that the result of the desacralization of nature in traditional Christian faith has been to desacralize both nature and human beings. "For most of the human story, nature was teeming with power, with life, and with an infinitely valuable order. These aspects of the experience of nature seemed to archaic men and women without question to represent disclosures of the presence of the sacred, to represent infinite powers of reality and of value …."

A growing number of theologians and scientists today are prone to blame Christianity for destroying pagan animism and thus destroying our reverence for the Earth as our mother. This criticism is badly misplaced. Christianity itself is a result of the destruction of paganism by agriculture. Agriculture brought a sense of control over nature, and with control came ownership and the loss of reverence. Nature became a commodity to be bought and sold, and the Earth itself was there to be exploited, even if it meant despoiling it. The sense of ownership which came with agriculture is central to Christian doctrine. The Earth and everything in it is ours. We don't belong to the Earth, it belongs to us. Christianity is a religion developed by and for agricultural people, and it reflects the agriculturalist's sense of ownership and entitlement regarding nature.

Hunter-gatherers never thought in terms of ownership of the land. To them the land and everything on it was like the sky, and just as one could not own the sky, so with the land. They did have a home range which they moved about in and protected against incursion by other groups if necessary, but they did not treat the land as a commodity which could be owned or bought and sold. Farming changed all that. The constant movement from place to place in search of game ended, and people stayed put on their plot of ground. This they protected against all comers, human or otherwise. Their survival now depended upon their ability to make that piece of land productive and to protect it. As Daniel Quinn has pointed out, this meant going to war with nature. Suddenly the same nature which had been a beneficent mother became an enemy that attacked their fields with weeds, insects, hail, drought, floods, rodents, and disease.

Our agricultural ancestors became the first and only species which systematically made war on the living things around them. As hunter-gatherers they would avoid large predators when possible and kill them only when necessary to protect themselves. Most in fact admired and revered the large animals for their strength and prowess. As farmers who lived their whole lives in one area they would hunt down and kill every large predator in the area in an effort, usually successful, to exterminate them so that they could not kill their livestock or threaten their families. We still engage in this kind of warfare today.

Life for the farmer became an unending fight against the enemies of his crops. An entire season's work could be destroyed in a few hours if a swarm of locusts swooped down and devoured his grain. Rodents could come in from the fields and eat the grain in storage, and large herbivores like deer could eat and trample the grain. Nature had become a malevolent enemy which had to be destroyed. Compared to the relatively slow-paced life of successful hunter-gatherers, agriculturalists had to rise before sunrise and work until after dark. It was back-breaking work, and the low protein diet of grain caused them to be of short stature compared to the taller well-built hunter-gatherers. They had worn teeth from eating course grain and bent backs from toiling in the fields. Their bodies were broken down at 30 and they were dead soon after.

> *And unto Adam he said,…cursed is the ground for thy sake; in sorrow shalt thou eat of it all the days of thy life: Thorns also and thistles shall it bring forth to thee; and thou shalt eat the herb of the field; In the sweat of thy face shalt thou eat bread, till thou return unto the ground..*
> - Genesis 4:17-19

Hunter-gatherers revered nature because they saw it as a cornucopia which supplied all of their needs. It was a Garden of Eden, a garden which they depended upon and to which they belonged. Agriculturalists, on the other hand, were dependent for their survival upon their own gardens, and they had to fight constantly to protect and maintain them against the forces of nature. Conquer and subdue is an agricultural concept. Nature had become the enemy. Christianity grew out of this cultural milieu but did not create it. All of the world's major religions are the products of agrarian cultures even though they have retained many of the practices and beliefs of earlier hunter-gathering societies. The mysticism, worship of idols, belief in omens and spirits – all came from the oral traditions and practices of pre-agricultural peoples.

Technological development accelerated cultural change by increasing food production and reducing the percent of the population needed to produce food. The invention of the plow 7000 years ago by Egyptians is probably the most significant technological achievement in human history in terms of influence on culture and the Earth. An acre of land which is farmed can feed up to 100 times more people than it can support by hunting and gathering. Writing, brick-making, the wheel, weaving, pottery making, and irrigation are all the products of ancient agrarian societies.

We no longer live in an agrarian society. Farmers make up only 2% of the U.S. population today and most Americans have never been on a farm of any kind. Agrarian religions are as out of date today as the hunter-gatherers mythology was to agriculturalists. The subordinated role of women in Christianity, Judaism, and Islam today are prime examples. The wives of ancient farmers were practically slaves, and girl babies were looked upon with regret and disgust. Boys on the other hand were viewed as capital since they could work the land and were the inheritors of family land. It was the sons who were expected to take care of parents when they were too old to care for themselves. A farmer with many sons was wealthy because he could grow more crops and hire his sons out to farmers unfortunate enough to have only daughters, and he knew that his sons would care for him in his old age. The major religions continue to perpetuate this male bias through denying women equal opportunity in the life of the church.

I have always marveled that social structures and behavioral guidelines developed by ancient agrarian and pastoral societies have scaled up as well as they have, but today's world is a powder keg waiting to go off. The discrepancy between the wealthy and the abject poor, between societies still trapped in the Middle Ages and those with highly developed technologies, between developed and developing nations, and the increasing competition for resources, are creating cultural stresses like those that build up in the Earth before an earthquake. Unless these stresses can be released, they will explode as suddenly and violently as an earthquake.

Where do the world's religions stand in all this? In the U.S. when we should be making an all out national effort to solve the problem of poverty and making efforts to bring the world's people together, we spend our time and energy fighting over whether evolution should be taught in the public schools, whether gays should be allowed to marry and raise children, whether brain-dead people should be kept alive indefinitely (and spending millions of dollars to do so while millions of healthy people live in abject poverty and have no access to medical treatment); spend billions every year supporting missionaries and religious predators; allow self-serving politicians to increase the wealth of the already wealthy at the expense of the poor and middle classes, and build mega-churches that are opulent by any standard of comparison. And all of this maladaptive behavior is in the name of religion.

The environmental crisis is really a crisis of the human spirit. We have lost our sense of kinship with nature. We will not solve the environmental tragedies facing us until we regain something of the true reverence for nature that the ancients had. Trying to save nature out of a sense of guilt, or because we feel responsible for it is not enough. The sacrifices we will have to make require much more. We have lost our fear of nature at a time when we have more reason to fear than at any other time in the history of our species. Natural systems do not gradually decline away to nothing when they are stressed; they decline to a certain critical point then collapse suddenly and completely. No one really knows or can predict what the long term effects of our attacks on natural systems will be or when the critical point will be reached which will lead to a catastrophic collapse.

When Boeing's 747 passenger plane was under development the engineers wanted to know how far the wings could be bent under pressure before they would fail. During a test a wing was put under greater and greater pressure. The engineers held their breath as the wing bent further upward than they had predicted. All at once it snapped with an ear-splitting sound. Natural systems are like the wing; when stressed they don't slowly fail; they resist changes until they can do so no longer, then fail catastrophically. By pouring our pollutants into the air, water and soil, destroying species of plants and animals at a rate that exceeds even the worst mass extinctions of the past, allowing agricultural land to be degraded and developed, and producing more people each year, we are stressing the resources and life support systems of the planet in an uncontrolled experiment to see how far we can go before one or more of the resources or systems fails. This behavior is not only rash, it is irresponsible in the extreme.

Modern science has explicated what archaic religion knew well: our utter dependence on nature for our being and for all the attributes of that being.
- Langdon Gilkey

Environmental concerns are at their root spiritual problems. The environmental dilemma is an outward manifestation of a crisis of mind and spirit. There could be no greater error than to believe it is concerned only with population growth, resource depletion and pollution. These are involved, but the real crisis is the failure to develop realistic human values which will enable us to achieve a sustainable Earth. Science can sound the alarm about approaching problems and provide solutions to those problems, but unless society at large has the will to make the changes and sacrifices necessary to achieve them, nothing will be done.

> *A truly ecological view of the world has religious overtones.*
> - Rene Dubos.

The field of ecology is one of the youngest of the biological sciences. Ecological research has given us a new understanding of our relationship to the Earth and spawned the environmental movement in the United States and elsewhere. Ecological studies focus on relationships rather than mechanisms, and they have helped to define a code of ethics regarding the natural world. Environmental ethics force us to look beyond ourselves as a species: to consider what is just – what is ethical – what is morally correct, to consider our spiritual relationships and obligations to the rest of God's creation. This, then, is a place where theology and science can find common ground, a place where bridges between science and religion can be formed.

As Daniel Kozlovsky has pointed out our efforts to clean the air and waters, to recycle, to cut back on our use of energy resources, to protect wildlife in national parks and preserves, to reduce the cutting of tropical forests have merely prolonged the day when we will face an environmental collapse. We cannot solve the environmental dilemma unless we are willing to wrestle with and resolve questions like; does the Earth exist only for the benefit of humanity? Do nonhuman species have an intrinsic or God-given right to exist? Should we consider the needs of future generations in our treatment of the Earth? Does our preeminent position among other life forms give us an obligation to be wise stewards of the land? How do we define our rights as members of an integrated web of life in which each part is dependent upon the others?

We know what must be done. There is no lack of information on how to solve environmental problems. The problem is our inability to develop values that respect the natural world for what it is: the source of all life, including our own. Deep within our Stone Age mentality, at the very core of our being, we resonate with everything that lives, and with the mountains and the seas and the celestial bodies. They are a part of us which the thin veneer of 12,000 years of agricultural civilization cannot erase. Denying who and what we are keeps us from coming to grips with the true nature of environmental and social problems: our inability to find our way back to those values which enabled our species to survive against all odds for more than two hundred thousand years.

James R. Curry

Science cannot help us find our way back, but religion can and must. I contend that saving the planet from degradation needs to take precedence over efforts to proselytize and accumulate wealth and power. It is not enough that prominent theologians are speaking out in favor of a life based upon environmental ethics; the effort must reach down to local churches and become a grass roots groundswell that will change the way people act and think.

What does the future hold for us as a species? I believe we are at a crossroads: one road leads to continued degradation on a global scale – a scaled up Rapa Nui - the other leads to reconciliation between races, religions, and nations, the development of sustainable lifestyles, and a greater respect for the other living things that share the planet with us. Changes of this magnitude will not come without a world-wide revolution, and revolution will not come without some massive discontinuity which shakes all of the peoples of the world to their very souls. I will not speculate on what the discontinuity will be – it could be a nuclear holocaust, a viral pandemic of unprecedented proportions, famine, or a series of natural disasters on a scale that defies imagination – but it will come, and it will forever change the way we look at ourselves and the natural world.

> *"Many ecologists....believe that we have one, at most two generations in which to reinvent ourselves and turn present trends around so as to prevent major catastrophes from occurring on Earth.*
> Harry Hollis

Religion

As a biologist, I cannot help but recognize the enormously important role religion plays and has played in one form or another in the life of all human cultures as far back as we can see. I am convinced that there is a biological imperative for spirituality that is built into our genetic makeup. The six chapters which follow explore the nature of religion today, how it relates to science, and the role it is playing and can play in helping all the peoples of the world achieve a deeper personal spirituality as well as global harmony and cooperation.

CHAPTER 21

A Living Faith

No one can say with any certainty just when the members of our species began to become aware of a reality that could not be experienced by the senses, but in all likelihood it occurred very early in our evolution and was a key element in our survival as a species. Even today, peoples stuck in a Stone Age culture live in a world of spirits and deities, and devote much time, energy, and resources to their worship. These are not isolated cases. Religion, in one form or another, is a universal characteristic of our species. It defines us biologically as much as do our unique hands or large brain.

Science has weakened the influence of the church by developing paradigms of the world based upon purely mechanical, physical, and chemical properties. This has encouraged an outlook that devalues and diminishes religious and spiritual values. Many of the strong spiritual ties that were so important in earlier human cultures are weakened or lost today. In an increasingly scientific and technologically minded society, spirituality is losing its relevance for a majority of people. I don't see this as a defect of science, but as a failure of organized religions to be open to growth and development. Our religious heritage mires us in ancient beliefs and practices, none of which have anything to do with our innate spirituality. We have allowed symbols and rituals to replace a living faith.

In a society that values individuality and independence, both science and religion elicit apprehension. Science, with its mechanistic approach, diminishes our humanity and seems to reduce us to the level of genetically programmed biological robots. Religion, with its call to faith inculcates an attitude that it is better to believe than to know, also diminishes our humanity by making us subservient to tradition and dogma. Restrained by ancient scriptures, rites, icons, and symbols that date back to our animistic roots, and a belief that all are holy and therefore nothing should change, organized religion today is on the defensive. Male dominated and female suppressed, it is not really people oriented, although it purports to be. People are not free to express their own spirituality but are confined by a more or less inflexible authoritarian system of codified beliefs and rituals that makes them spectators in an area of human activity where they should be full participants.

At the same time, both religion and science empower us and help us to deal with a universe that seems uncaring and hostile. Rene Dubos, a molecular biologist and two-time Nobel laureate, in his book *The God Within*, had this to say about the relationship between religion and science:

> *Religion and science...constitute deep-rooted and ancient efforts to find richer experiences and deeper meaning than are found in the ordinary biological and social satisfactions... both the myths of religion and the laws of science...are not so much descriptions of facts as symbolic expressions of cosmic truths. These truths may always remain beyond human understanding, but at every stage of human development glimpses of them have enriched man in experience and comprehension.*

Both science and religion hold out the promise of a deeper, richer, more fulfilling life, and both have failed to provide it in modern society. We are capable of so much more than our materialistic society allows. All of us are trapped in a culture to which both science and religion have contributed major components, a culture which promotes and values material success and power over spirituality and interpersonal relationships. Science has achieved what organized religion has striven for but never achieved: unification. Science speaks with a single universal voice and a common language. As a way of knowing, science has been far more successful and *intellectually* satisfying than religion. Scientists universally accept the same broad tenets regarding the natural world, while at the same time being open to changes in the basic constructs of their respective disciplines. When scientists meet or correspond political differences are irrelevant, as are religious or cultural differences. All speak the same language - the language of science. Communication between scientists is free and open, and disagreement is not only encouraged but seen as vital in the pursuit of new knowledge. This promotes creativity and imagination which leads to new insights into the workings of the universe. The result is a continuously expanding scientific world view that can be breathtaking in its scope. Unification gives science and scientists an authority, power, and prestige beyond political boundaries and cultures which theologians and church leaders only rarely enjoy.

Organized religion, on the other hand, speaks with a Babel of tongues, and fiercely defends its own dogmatic precepts, often to the death. Ancient beliefs, some of them undoubtedly reaching back to the Stone Age or beyond, are treasured and inviolate. Change, when it comes, is painful and lags well behind need. Creativity and imagination are discouraged because they threaten the status quo. The natural curiosity and independence of our species is stifled, and believing is more important than knowing.

Science and religion both empower us, but religion reaches deeper into the human psyche and enables us to supersede our weaknesses and limitations, to go beyond what we normally feel capable of suffering or achieving. Our beliefs help us to accept tragedy that is beyond accepting and lifts our spirit to heights unattainable in any other context. Together, religion and science have the potential to give meaning and substance to our lives that cannot be achieved by either acting alone.

Religion and science should be servants, not masters. Traditions that do not serve people in their everyday lives should not be tolerated. This is the golden mantra of science and should be of religion. Both religion and science are not only failing to resolve the major social problems of our time, they are exacerbating them. Far too much in the theology and rites of today's religions are little more than a house of cards – layer upon layer of mythology which cannot be sustained rationally, by experience, or by the needs of our lives . We accept that house of cards because our fathers did.

Beware of new things, for every new thing is an innovation and every innovation is a mistake.
- Mohammed

In her book, *The Trouble With Islam*, Irshad Manji contends that Arabs have no sense of future because of past failures; consequently, she says they look only to the past. She calls this foundamentalism. "As Muslims have slipped further behind Europeans in military and material honor, momentum has gathered for foundamentalism. The entire exercise of washing prescribed areas of the body, reciting specified versus and prostrating at a nonnegotiable angle, all at assigned times of the day, can degenerate into mindless submission and habitual submissiveness." Mindless submission and habitual submissiveness is not a characteristic of Islam alone. There is much in Christianity and Judaism that also denigrates into mindless submission and submissiveness, but if formal religion can become encrusted and rigid, highly personal religion, which Pope Benedict has warned against, can suffer from a lack of institutionalization. The social aspects of religion are important in developing and maintaining group identity and in reaffirming religious beliefs and commitments.

I have never had a religious experience, one that would change the course of my life, but I lived with an extended family for 21 years who had such experiences and who had an unshakeable faith in God. The whole texture of their lives was woven around the Church and their religious beliefs. They were (and are) intelligent and honest people, not easily misled by false hope or the rationalization of life's problems. On the contrary, their faith enabled them to face challenges, no matter how difficult or tragic, with a self-assurance and forthrightness that made ours a secure and loving family.

My experience is not unique. Hundreds of millions of people of all races have found similar strength and comfort in faith without sacrificing their own dogged determination to confront life directly and work their own way through the hard spots. True faith is not weakness: it is strength of a special sort. It is the strength of self-assurance; it gives meaning and value to pain and suffering in a way that nothing else can provide . And it gives hope; not the false, wishful kind of hope that attends the buying of a lottery ticket, but a genuine expectation that life has meaning and value.

Ridiculing such strongly held beliefs, as a minority of scientists are wont to do, is an attack on our humanity. Beliefs like these are ancient, and so universally held as to reveal a need and a solution that is programmed into our nature. A belief in the supernatural is as much a species-specific characteristic as language, technology, and cultural evolution. To my knowledge, there has never been a society or culture that did not actively search for meaning, as well as structure, in life and in the universe.

Our ancient predecessors devoted enormous amounts of time and energy in their efforts to communicate and placate the supernatural. This was time and energy which could have been devoted to acquiring food, constructing tools and buildings, and other activities that offered a more direct and tangible reward. Enormous expenditures of time and energy which do not bring about a payback in survival terms is a certain and short pathway to extinction. Burial rites, decoration of clothing and household objects, music, art, the construction of altars and icons, all have been intimately associated with and developed through religious beliefs, and none of these activities directly provides more food or protects from predators, disease, or accidents.

How then, in scientific terms, can the survival of such beliefs be explained? I don't have all the answers, but one thing seems clear – religious faith and the activities it spawns are now and always have been integral to our survival as a species. Ignoring or ridiculing this basic aspect of our makeup is a mistake too frequently made by people who look at life and the universe only in mechanistic terms. There is a creative force in the universe that cannot be ignored, and our evolution as a species has imbued us with the intuitive capacity to respond to it.

CHAPTER 22

Love Is Eternal

God's people never say goodbye for the last time.

- Message on a church sign

In his award winning film, *People of the Forest*, Peter Van Lawick chronicles the life of a female chimpanzee named Flo and her four offspring over a period of 20 years. The two oldest siblings were males, the third was a female, and the youngest was a male named Flint. Born late in life to the mother, Flint was, in the words of Van Lawick, a brat. The aged mother was unable to keep up with or discipline the hyperactive infant who terrorized his siblings and other members of the group with his aggressive antics. Flo did not permit any of the others to raise a hand to the young chimp, and at times he got out of control and caused a ruckus in the entire troop.

Flo died suddenly of heart failure when the youngster was 8 years old, and her body lay alongside a small creek. She was 53 years old. The whole family gathered by her body and other members of the troop slowly arrived, as if to say goodbye to their friend. Flint was distraught and would not leave his mother's body. He lay down close to her, frequently touching her and brushing away flies. One by one the other chimpanzees drifted away, and finally Flo's two sons and daughter went with them into the forest, but Flint would not leave. He climbed a tree overlooking the creek and made a nest. He remained by his mother's body for three weeks before dying. Van Lawick concluded that Flint died of grief. I have no reason to think him wrong.

James R. Curry

I change, but I cannot die.
- Shelley, The Cloud.

We all know the enormous sense of loss that accompanies the death of a loved one. The finality of it is overwhelming. We feel confused and isolated. People and events move around us as if in a dream. It is at times of deep grief that most people fall back on their faith for comfort and the will to go on with their lives. Our loved ones are not really dead; only the body is dead. The mind, the spirit lives on in a far better place and we will someday be reunited.

But like little Flint, some people are unable to accept the finality of death, and take their own lives or spend the rest of their life pining over or memorializing the lost loved one. This is most common when children are lost. A sixteen year old young man in my community died suddenly during a football practice. His father was never able to accept the loss. He built a huge limestone memorial at the gravesite and visited it every day for years, sitting for hours by his son's grave. Such grief is not uncommon, although it may not always be expressed in this way. The man's open expression of grief may actually be healthier than bottling it up inside where it can erode the will to live.

The death of a loved one is the greatest discontinuity any of us ever has to face. Death is difficult to accept because our minds require order, even if it is imposed and has no basis in reality: tree, dog, storm, mountain – a myriad of objects – we identify them and unify them into larger mental constructs – valley, lake, ecosystem, world, universe, none of which have any reality other than our idealized concept. Not only is each entity unique, but each is constantly changing in time and space. Nonetheless, without such mental constructs we would be lost in a chaotic world without form or meaning.

Our minds have difficulty dealing with the chaos produced by a lack of order. It can drive us insane. Mass hysteria would result if the sun were to rise in the west some morning and set in the east. A startling discontinuity like this would cause some people to turn to scientists for an explanation while others turned to their religious leaders. When neither could offer acceptable explanations, how would people react? I believe that millions would commit suicide rather than try to come to terms with the unexplained phenomenon.

One brilliant summer afternoon when I was a child it began to grow dark in mid-afternoon. There was no explanation for what was happening. People all up and down the street came out of their homes and stared up at the sky. They gathered in small groups and spoke in whispers, and there was an atmosphere of foreboding. Although I was a small child at the time, I still remember clearly the strained look on peoples' faces as they tried to comprehend what was happening. A few seemed to be praying silently while others crossed themselves. The episode lasted through the afternoon and evening. The next morning the sun rose bright and clear. We later learned that smoke from a prairie fire in Kansas was responsible, but what if there had been no communication system to tell us the cause?

How would a primitive tribe have reacted to this unknown and unexplainable discontinuity? Chances are they would have been terrified and would have called on their gods to return the sun to them. When it returned a few hours later, they would have believed that the gods heard their supplications. Science and technology explained the phenomenon to the people of my neighborhood; a primitive people would have resorted to religion. In both cases, the fear of the unexplained would be relieved, but in some ways the primitives would get the better of it because their faith in their ability to influence the gods would be reinforced. That would give them a sense of security that science and technology cannot give.

There must be order beyond what our minds are capable of imposing on the natural world. Religion is a way to explain the unexplainable and to endure the unacceptable discontinuities. God is an all-embracing mental construct for imposing order on the universe that is one very large step beyond science.

Ancient peoples often tried to recreate the dead person's life by placing in the tomb objects which had been important to the deceased in life, or representations of such objects. The ancient Egyptians decorated the tombs of important people with wall paintings and miniature people, animals, boats, plants and other objects. Loren Eiseley commented on this practice in "Days of a Thinker:"

> *Men of power in the Nile kingdom believed with a touching simplicity that you could take everything with you in miniature: your little fish pond of hammered copper, that would hold water, your spotted cattle, your pleasure boats, musicians, serving girls, even yourself in effigy. It was a gentle game, a game of childlike make-believe. There were wood carvers whose duty it was to create the pleasant palace and the precious palms, everything to insure that life would go one with no loss of status, that there would be an oarsman at the pleasure boat throughout eternity. I thought...how enticing it would be if they truly believed, there in civilization's dawn. If the funeral toys really comforted, if we could still, in some magical manner, be transported to where little figures in wood forever defeated the sting of death. I wondered, grubby student among old books, if they genuinely meant it, if they had known*

a secret I had tried in a cruder, unskilled way to reproduce with gilded crosses. How similar we all were across the millennia.

There is no greater discontinuity than death. Death is the end of mind. Nobody cares about the death of the body; it is the mind and the mental constructs that have developed over a lifetime that matter. What will it be like the morning after my death? Will the rain still fall, trees still bend to the wind, birds still sing, the world still exist? It is incomprehensible that the mind should cease to exist, and so we cling to life even in great suffering, for suffering is also a manifestation of mind.

The concept of a soul and a God allows us to order death. The mind will not only still exist, but it will be a higher level of existence where all discontinuities will cease and peace will reign forever. The mind will be free of the limitations of the mortal body. All pain, frustration, illness, and emotions will be gone. It is a perfect mental order of a kind not possible in this life. Whether it is true or not doesn't matter – it works. It helps us to accept not only our own deaths, but the deaths of loved ones. They have gone to a better place, but mind still exists, and when we meet them in the hereafter they will recognize us, remember our names and the experiences we shared together while on earth. Mind can walk away from body the way we walk away from the hair left on a barbershop floor.

Deep down we may recognize, as perhaps the ancient Egyptians did, the inadequacy and artificiality of our mental constructs and wish for better ones. Some people turn to science for help, but science can provide it only in limited amounts and is likely to fail us when we need it most. Religion is the best avenue for what we seek. It accepts the fact that in our present state we are incapable of a perfect relationship with our world or with God. Science holds out the unstated promise that given time we can understand the natural world so thoroughly as to gain complete control. Indeed, science has been so fruitful as a way of knowing, that society at large, including many modern theologians, seem to believe that science holds the key to humanity's needs and desires. This is rooted in the belief that a perfect natural order exists, a belief which has it origins not in science, but in philosophy and theology.

Burial

A solemn group of about twenty people made their way up the side of the mountain and into a small cave. All wore their best clothes and adornments. The women wore colorful flowers in their hair. When someone spoke it was in a whisper, and a few people sobbed quietly. The open grave about which they clustered was dark with shadows and the bottom could not be seen. All eyes turned toward the mouth of the cave as six men approached carrying the body. A shaman preceded them speaking slowly the words of the death chant and rattling a string of bones.

The body was laid on a bier of branches and each person in turn went to it and stood for a few minutes, gazing into the face of a man they had known for most of their life. When all had said their goodbyes, the body was tenderly carried to the grave and placed inside in a flexed manner with the knees bent to the chest. A few of the dead man's prized tools, some food, and bunches of colorful flowers were placed in the grave. Each of the people in attendance then threw in several hands full of dirt, and the grave was filled in by the elders. Slowly the group moved out of the cave, but a few lingered, not wanting to leave their friend and leader.

The story is very familiar to all of us. It is a measure of the universality of grief that ceremonial burial is characteristic of all human societies. This particular burial, however, was not of a modern human. It occurred 60,000 years ago in northern Iraq by members of the human species we know as Neanderthal. The skeleton was discovered in 1960 by Ralph Solecki in a cave called Shanidar. Soil samples collected from the grave contained thousands of pollen grain from eight species of brightly colored spring flowers. Grape hyacinth, bachelor buttons, hollyhock, rose mallow, and others were included, all of which grew on the mountain where the cave was located but which were widely dispersed. Some of the pollen inside the pit was still inside the pods that produced it and could not have been blown into the cave by the wind. Solecki concluded that the extreme concentration of pollen in the grave indicated that bunches of flowers were placed there at the time of burial. The grave at Shanidar is the earliest documented instance of ceremonial burial in a humanoid species.

Other Neanderthal burial sites also show evidence of ritual burial which includes the inclusion of tools and food. It would appear that Neaderthal had an awareness of their own mortality and mourned their dead. That is not too surprising given that these humanoids built shelters, controlled fire, skinned animals, made tools, and shared many other characteristics with our species.

Burial in today's world is a uniquely human activity, and for the most part, it is associated with the belief in an afterlife. Does the fact that Neaderthal buried their dead imply a belief in an afterlife? There can be other explanations for burial: getting rid of a corpse that would attract dangerous predators and scavengers; eliminating a source of foul odor; hiding from enemies the fact that a leader has died; but none of these reasons could account for a ceremonial burial and formalized ways of placing the body in the grave. The evidence is too scanty to conclude that Neanderthal had a belief in an afterlife which would signify the beginnings of religion, but clearly the uniquely human act of ritual burial raises the question.

We are not the only animals to grieve our dead. The story of Flo and her youngest son at the beginning of this chapter is evidence for grief in other animals. Elephants are known to return periodically to the skeleton of relatives. They are very quiet as they gather around the bones and gently touch them with their sensitive trunks. E.O. Wilson has said that love is an old mammalian social trait. Birds, fish, amphibians and reptiles are often very protective of their offspring and mates, but the close social bond that we call love is absent or very weak in them. I have kept a green iguana for nearly twenty years, fed him regularly, and cleaned his large cage. He likes to be touched and will close his eyes and raise his head in obvious pleasure when I stroke him, but he does not respond to me with any show of affection or even recognition. The same is true of pet birds, frogs, or fish. On the other hand, dogs, cats, pigs, rats, and most other mammals show a great deal of affection when raised as pets.

> *Set me as a seal upon thine heart,*
> *As a seal upon thine arm:*
> *for love **is** strong as death* - Solomon's Song 8:6

Deep grief is borne of love, and love does not end with the grave. We mourn our lost loved ones forever. Time will soften the loss, but not erase it. With the loss of a loved one a door closes to a room in our psyche that is never reopened. We may walk past it and pause before the door, but we can never enter. The inability to accept the finality of death and the realization of our own mortality were probably the original bases for a belief in an afterlife. If little Flint could have believed that he would see his mother again in an afterlife, he may have been able to accept her death and go on with his life. Being incapable of that he chose to die rather than leave her.

> *Happy birthday in Heaven, Mom.*
> *I thought of you with love today,*
> *but that is nothing new.*
> *I thought about you yesterday*
> *and the days before that too.*
> *I think of you in silence,*
> *and often speak your name.*
> *Now all I have is memories,*
> *and your picture in a frame.*
> *Your memory is my keep sake,*

with which I'll never part.
God has you in His keeping
I have you in my heart...
Daughter's memorial to her mother in the newspaper.

It broke our hearts to lose her,
But she did not go alone,
For a part of us went with her,
The day you called her home.
Our hearts ache with lonliness,
Our eyes shed many tears,
And God, you know
How much we miss her...
At the close of ten sad years.

Parents newspaper memorial for their daughter

Touching memorials like these can be found periodically in every newspaper in the country. They express the sadness and everlasting longing for lost loved ones, and a majority of them express the belief that they will be reunited with the loved one in Heaven.

Ancestor Worship

My father smoked cigarettes from the time he was twelve years old until he was sixty. The result was emphysema and early retirement at age 62. Several years later he was hospitalized with a collapsed lung, although at the time his doctors were unable to diagnose the problem. He went into a coma and was near death. Once the correct diagnosis was made and treatment begun he began to improve, much to the relief of the family members who had gathered at his bedside.

I remember vividly his waking up and looking at me for long seconds. He appeared to be confused at first, but slowly shook off the effects of the coma. When he finally spoke it became clear that he knew he had been near death, in fact that he thought he was dying. Tears filled his eyes as he related in a weak voice how during the coma his mother's face came to him and brought him peace. In describing the effect this had on him he used words like powerful, reassuring, but it was clear as he struggled for words that there were no words which could adequately describe the experience. It had been so real, as if his mother had actually come to him. That she had been dead for decades did not matter in the least. She came to him when he needed her, and the effect was to bring him back from the grave. He lived another twelve years.

We all know how real dreams can be and how they can influence our lives. While asleep we can live our fantasies or walk the corridors of time to meet loved ones long gone. I sometimes regret waking up from a particularly pleasant dream. Some people, and my father was one of them, believe that God speaks to us in dreams. He sometimes would relate a dream to me and explain what the Lord was telling him. The problem with this was that I didn't always agree with the interpretation. Dreams are like ink blots: interpreting them is a subjective and purely personal thing. That in no way discourages the faithful from their belief; in fact, it reinforces it because they believe that God is speaking only to them, which makes it very personal indeed.

Perhaps it was dreams which caused some cultures to begin worshiping their dead ancestors. The worshiping of ancestors continues today in parts of Africa, Asia, and the South Pacific. Ancestor worshipers believe that the spirits of the dead continue to dwell in the natural world and that these spirits are always nearby, observing the lives of their descendents. Because they have supernatural powers and can either aid or harm the living, they must be venerated and placated through various rites and practices. The following excerpts from a third year writing student was posted on http:/virtualdoug. typepad.com/virtualdoug/2005/05/student_essay_v.html

> "Brought up in a traditional Vietnamese home, I have been taught that ancestors are profoundly important because their spirits are watching over all of the family's members. Like other Vietnamese people, my parents show their deep respect for the ancestors by setting worship altars in our house and spending time taking care of it everyday. Generally speaking, Vietnamese think that in death, one does not pass away. Instead, one passes on to another world, which invisibly exists beside the land of the living. The dead people whom no worship is given, are disturbed in death and prey on the living. Therefore, Vietnamese consider ancestor-worship their obligatory duty. In addition, Vietnamese believe in the supernatural powers of the dead people, which can bring them happiness, good luck and even money.

Ancestor worship implies a strong belief in life after death. Herbert Spencer considered it to be the earliest and most rudimentary form of religion, but this is probably going too far. Still, one has to wonder at what point and in what way in the course of our evolution we began to identify and value self to the point that we have difficulty accepting death. The will to live is paramount in all species or they could not survive, but is it the result of sentience, or simply a computer-like genetic imperative?

When we near the end of our life and are forced to face our own mortality, where do we turn for solace and courage? Having lived a highly socialized life, we are now forced to face the most threatening challenge of our lives completely alone. No one can walk through that dark door with us or for us.

> *Precious Lord, take my hand*
> *Lead me on, let me stand*
> *I am tired, I am weak, I am worn*
> *Through the storm, through the night*
> *Lead me on to the light*
> *Take my hand precious Lord, lead me home.*

When all of the electronic equipment humming at our bedside and all of the drugs dripping into our veins have failed, where do we turn for help in accepting our end? For many people words like those in Thomas Dorsey's beautiful hymn provide an answer. Religion can take us where science and technology cannot; in fact, science and technology are most likely to fail us when we need them most. When loved ones die, when serious illness overtakes us, when our families fall apart, when accidents strike, when we find ourselves alone in an uncaring world, when financial crises befalls us, it is often our spiritual strength that gets us through the day.

There are also public outpourings of faith following tragedies such as the Oklahoma City bombings, 9-11, and natural disasters like the hurricane which devastated New Orleans in 2005. Passengers on a flight in the Hawaiian Islands some years ago began singing hymns after most of the roof of the airplane fell away and they were looking up at the sky expecting to die. Legend has it that the band on the Titanic continued to play hymns as the ship was sinking. Not everyone feels the need to turn to expressions of faith at times of crises, but even people who proclaim to be agnostics or atheists will sometimes do so.

If little Flint had been fortunate enough to have a faith to turn to when his mother died, perhaps he could have accepted her loss. One can only wonder what went on inside his mind. Did he have a realization of what death meant? Could he understand the finality of the loss, or did he linger near her body in the hope that she was just sleeping and would awaken to love and protect him once more? We will never know.

James R. Curry

> When I have known the last of all tomorrows
> And torn the petals from one final rose,
> Then I shall go as quiet water goes,
> Upon a certain way I knew and chose.
> And the gray, dragging hours I shall forget,
> Exploring worlds I know not yet.
> Lois Anne Davison

CHAPTER 23

The God Within

There is strong archaeological evidence to show that with the birth of human consciousness there was born, like a twin, the impulse to transcend it.

- Alan McGlashan

Religion has always been the most powerful force in human culture. Religion satisfies our most basic needs through ritualizing art, music, and poetic speech. It fosters social interaction and a sense of belonging to a community. Religion helps us soften our egotism, bear suffering, manage our guilt, and persevere through grief. It mutes our fears and gives us hope.

The roots of our religious nature are old – older than science, older than mankind. Science is cultural and but a few thousand years old; religion is primal. Modern human culture began about twelve thousand years ago with the introduction of agriculture, but our species is hundreds of thousands of years old.

I believe we first saw God through our hands, for with hands came the ability to create, and with creativity came imagination. The hominids that preceded us could pick up a piece of formless stone and see within it a spear head or hand axe and could imagine its shape, size, weight and how well the implement would perform the task for which it was designed. As creators, they could marvel at the natural creations which make up our world and imagine that they too must have a creator, and that just as our spirits live in our creations, so the spirits of the natural creators live in theirs. Perhaps because their lives depended upon everything in the natural world, they began to worship the creators of that world as being far superior to themselves;

indeed, as being supernatural.

If only we could travel back in time for millions of years and look through Lucy's eyes for a moment, or through the eyes of Handy Man or Erect Man, we might know when the first glimmerings of worship began. They had our hands, and although their ability for abstract thought was not yet as well developed as ours, they were in every sense of the word, creators, and hence could appreciate to some degree the creations of nature.

It is easy to laugh at the beliefs of the ancients – that mountains had spirits, that the sun god could withhold his life-giving energy, that evil spirits were all around which could bewitch or harm - - but these beliefs served them well. Their beliefs forced them to analyze and appreciate the natural world in a way that contributed to their survival. It forced them to study natural events and see patterns in nature they could exploit. It was a way to delineate between what was useful and what was dangerous. Perhaps most importantly, it fed their imaginations, which stimulated new ideas and helped drive natural selection in the formation of a larger brain.

The ancients devoted much more time and energy to their worship than most people do today. That was energy that could have gone into procuring or producing food, making tools, or a hundred other activities which would have aided their survival and prosperity. That they chose to spend resources, time, and energy on worship is an indication that it contributed something very substantial to their survival. If worship had been a drain on resources and energy without making a contribution which exceeded the drain, the species would not have prospered and could have died out. Natural selection favors what works in the competition to survive: the fossil record abounds with species which disappeared because their energy books did not balance. Falling into the "red" in the natural world is the quickest way to ruin and extinction.

> *I am half pagan; goat gods of the Greeks,*
> *Satyrs and fauns, are charming to my eyes,*
> *And I can mock the stumbling heart that seeks*
> *Some vague all-father hid beyond the skies.*
> *Life in the earth is all I hope to hold,*
> *And fleshly temples outstrip those of stone,*
> *My soul is incense in a box of gold,*
> *My brain is nectar in a cup of bone.*
>
> *I cannot kneel with pious heart in prayer*
> *While satyr music haunts the wind-washed hill;*
> *With garlands decked I must be up and there;*

Like reeds in wind the calling pipes are shrill.

But when creating, I believe in God,
Like Him on the unloving breathe my breath;
Thus did He thrill when life moved in the sod;
With Him I mold and make, and know no death.

- Lois Anne Davison

What does religion do for me? Most people don't ask that question, at least not consciously. It is not an easy question to ask oneself, and an even more difficult one to answer. Many people would say that it helps keep faith with their parents. They raised me to believe, and I continue to believe so that the family tradition will go on. They might also add; I believe, or at least I continue to observe the rituals of religion, in order to avoid feeling guilty about abandoning family traditions. Clearly, such hollow beliefs will not be of much help in facing life's challenges. Beliefs and practices ingrained in us as children are often difficult or impossible to turn away from, even though they really no longer have any meaning in our lives.

We are a social animal in the extreme. Our survival throughout the ages has hinged on our ability to function in social groups. Church services are primarily social functions. We sing together, and the power of the group reaches deep inside and makes us feel a part of something greater than ourselves. We listen, or half listen, or don't listen at all, to the sermon and rituals, and for a short while at least, forget the problems that plague our lives. The church is a sanctuary from the outside world, a place to pause, reflect, and strengthen ourselves for life's daily battles.

But is this enough? Is this all we should ask of our religion, our church, synagogue, mosque or temple? Much of what we get in a religious service can also be gotten in other social venues like a movie theater or sports stadium. An engrossing movie or exciting sports spectacle offers the same kind of social togetherness as a religious service. We laugh, cry, and applaud together, and for a short while, we transcend ourselves and forget our troubles.

I am no longer afraid.

Most people live under a burden of fear. We don't parade it around like our ancestors did, with their magic incantations, painted faces, and ritual sacrifices, but it is there nonetheless under a thin patina of modern sensibility. In the dark of night the fear often takes hold of us and makes sleep impossible. Long and lonely are the nights ruled by fear. We can alleviate the fear with lights, alcohol, or drugs, and sleep fitfully if not soundly, but the underlying fears that are so magnified by the dark shadows of the night are always there, always burdening our every thought and action.

Our fears are not new to us; they are older than the first creatures to climb out of the ancient swamps onto dry land. Fear, like pain, is nature's way of protecting us from the very real dangers of our world, but the primal fears of our bestial progenitors are infinitely magnified in the expanded cerebral cortex of our enlarged brains. There, among billions of neurons crackling with creative energy, resides an imagination that turns our fears into a reality that becomes increasingly difficult to bear. We are able to anticipate, more so than any other creature, the events which will engender fear, and we brood about them and allow them to influence our lives. Although we are in the best of health, we worry about illness. Although we are young, we worry about becoming old and infirm. Although we are financially well off, we worry constantly about losing that which we have, and seek ever more in an effort to alleviate the fear. These anticipatory fears rob us of some of the enjoyment which our lives should afford, and for some of us, they become so real as to be unbearable. As a species we are unable to live only in the moment.

Among the many fears that burden us the greatest is the night from which we shall never return. Other animals die quickly and without forethought; we begin to die from the moment we come face to face with a realization of our own mortality. To some the realization comes with the death of a friend or loved one. To some it comes with a near death experience at work or on the highway. To some it comes with a near fatal illness. And to some it simply comes with increasing age. Once we realize the certainty of our own end, we never again sleep as soundly or enjoy life with the same innocent joy we had as children.

Imagination is the lifeblood of our culture and humanity. Einstein once said that imagination is more important than knowledge. Science, technology, religion, philosophy, music, art – all are the brainchildren of human imagination, and all have the power to ease our fears and bring at least temporary peace and joy, but the oldest and most universal of these is religion. Religion can also be the most effective in bringing us peace, for religion incorporates all the others except science, and only religion offers the promise of eternal life.

In his book, *Orisa*, Philip Niemark talks about his conversion to IFA, an ancient religion practiced by the Yoruba people of Nigeria. He states that "IFA provides a pragmatic and psychological cushion against the vicissitude of life." He goes on to say that IFA divination survived fairly intact its transportation to the new world with slaves because, uncommonly, it worked. "Its efficacy was not just in the appropriateness of its verse but in the accumulated power of the spoken word." When asked by a reporter - "What is the single most important thing that IFA has done for you?" he surprised himself by answering without hesitation, "**I am no longer afraid.**"

The power of those words almost overwhelmed me when I read them. I wondered if I could ever find a religion that so completely satisfies my need for comfort, reassurance, and a belief that my life has ultimate meaning that I could live without fear. And I wonder how many people of any religion can say truthfully that because of their religion they are no longer afraid?

Yea, though I walk through the valley of the shadow of death,
I will fear no evil: for thou art with me.

Much has been written recently about the increasing secularization of Europe. Pope Benedict XVI addressed this trend by warning Europeans against growing secularization and "do-it-yourself" religion. He told those attending the church's 20th World Youth Day festival that there is a "strange forgetfulness of God....religion constructed on a do-it-yourself basis cannot ultimately help us." Would people desert Christianity or any other religion if it provided a genuine cushion against the vicissitudes of life? I think not. People do not turn their backs on that which gives their lives meaning.

Clearly, increasing secularization proves that modern religions are failing to provide a real cushion against the demands and trials of modern life. The stubborn rigidity of religious tradition may have been meaningful to people in the past, but it is very much out of synchrony today with what many people need and want in their lives. Men hundreds of years dead still hold us in their icy grip. Religion is a fundamental intuition of mankind. It is basic to our nature. No religion reaches as deep as a personal faith, but for far too many people being religious simply means accepting a subscribed list of beliefs and joining a group. That so many people today are choosing to make religion a personal journey rather than a group-related one is indicative of the need for religion in peoples' lives as well as the failure of established religions to satisfy the need.

Most ways today of being religious actually retard, rather than promote spirituality. Far too many churches, synagogues, mosques, and temples are for the most part political and social entities that seek power, influence and contributions.

James R. Curry

Holy Places?

We apparently have an incredibly strong need to find something tangible which will confirm the correctness of our beliefs. We need, desperately, something we can see, touch, or walk on to feel closer to God. In this regard as in many others, we are no different than our ancient ancestors who wore skins and trembled before the spirits of their holy places. Following the American civil war many people believed that spirits of the deceased could be captured on photographic film. Photography was in its infancy then and seemed to have a magical quality about it. A photo of Mrs. Lincoln seated on a chair with a barely visible outline of her dead husband standing behind her with his hands on her shoulders was widely circulated as evidence that the dead watch over and care for us. These were, of course, double exposures produced by photographers for sale to the gullible, but the nation had recently lost over 600,000 men in the war, and there was a widespread, desperate need to believe that death was not the end of life.

Holiness is not a place; it is a mindset and a way of life. It is not about where I am, but about who and what I am or are to be. Deep feelings can be evoked by physical things. I will never forget my visit to the Viet Nam War Memorial in Washington, D.C. Others had told me how deeply they were moved by standing in front of the wall, but I did not expect to share their experience. As I walked along that black, shiny vertical surface with tens of thousands of names inscribed on it, I was overwhelmed with emotion in a way that I had never experienced before. Just writing about it now brings back some of those emotions. The enormity of the tragedy revealed by that wall was brought home to me in a way that would have been impossible otherwise.

Is the wall holy to me? No. Do I revere it for what it stands for: a monument to the young men who sacrificed their lives and a visible testament to the futility and stupidity of war? Yes. Would I kill other people to keep them from visiting or owning it? No. How holy can something be that engenders internecine warfare and the deaths of thousands? I'm sure that when a Muslim visits Mecca or a Jew stands before the Wailing Wall they experience deep emotional feelings similar to those I felt at the War memorial. But if they truly believe that these are in any way portals to the supernatural, how could they act in ways contrary to the values inspired by that supernatural? We have had several thousand years in which to progress beyond animism and the superstition upon which it is based. We should know better, but intelligent and creative people cling stubbornly to the concept and kill each other for possession of these supposed "holy" places. Fighting over "holy" places is barbaric and cannot be justified by the high ideals of religion. The world would be a better place if our concern was for humanity, not barren ground or other natural or man-made objects.

> *"May all who reject God live and die in misery."*
> ("Let it Out" Comments, Indianapolis Star, Nov. 12, 2005)

Extreme dogmatism is wrong whether practiced by scientists or fundamentalist of whatever religion. There is a significant difference between the two groups, however, because scientists belong to a culture that aggressively encourages dissent and change. Respect for differing viewpoints is a cardinal principle of the culture of science. When Darwin embraced a naturalistic view of the origin of the Earth and the living things on it, he was going against the natural theological views held for centuries by the overwhelming majority of scientists. When Alfred Wegener proposed the theory of plate tectonics, he was ridiculed in some quarters, but scientists tested his radical theory and found that it was correct. As a result, we now understand much about our planet and its violent geology than was unknown before. There is almost nothing in science that goes unchallenged. Every theory or hypothesis is tested again and again by the scientific community with the expectation that no theory or hypothesis is inviolate and will eventually change.

This is in stark contrast to the resistance to change which is found in most religious groups. If God is the same today, yesterday, and forever, then change of any kind is wrong. God is perfect, and perfection by definition is unchangeable. What He revealed to our ancestors millennia ago is just as valid today as it was then. If science operated in this way we would still believe that the Earth is the center of the universe.

Religion at its best brings out the highest qualities of humanity. Religion at its worst brings out the worst qualities of humanity. It is heaven and hell on earth. The same can be said for science. People everywhere long for peace, happiness, and security, but to what extent have any of these increased in the last 10,000 years? Neither religion nor science has moved us any closer to these ultimate human goals; indeed, both have been hindrances to their achievement: religion through violent intolerance, and science through the development of ever more violent means of killing each other. The failure of the two most powerful forces in our history to bring us any closer to universal peace and security than our stone-age ancestors enjoyed is an indictment of our humanity which cannot be explained by the concept of a benevolent creator, but which can be explained by the fact that we are 21st century creatures trapped in the genetic wrappings of stone-age men. The behavioral traits which served the ancients so well – fierce territoriality to protect priceless resources, domination of women by men, technological prowess for developing new tools and weapons, a strong group identity which excludes members of other groups, an aggressive spirit when dealing with threats, a fascination with sex – have prevented us from achieving the kind of lifestyle which every member of the human family so desperately wants. Our species is hundreds of thousands of years old, but modern culture is not much more than 10,000 years old. That is too short a time for our genetic makeup to adapt to our changed circumstances.

Religion and science can be the most empowering, most liberating, most life-giving forces in our culture, but all too often they have been a tool for greed, manipulation, and violence. Both have failed miserably to substantially better the human condition, but the greatest failure has been that of religion. Science historically has never dealt with values. That is changing today, particularly in the environmental sciences, but scientists have always looked at the world and universe in purely rational, objective terms.

While science and technology have supplied the weapons by which we kill each other religion has always supplied the justification. This is particularly true of the monotheistic religions. The belief in one exclusive all-knowing god is fundamentally intolerant. All other gods are by definition false and anyone straying from the approved church doctrine is a heretic. Since God's purposes are inscrutable to mere mortals they must be obeyed without question. The atrocities committed in the name of Christianity provide a litany of vengeance and barbarity almost too painful to recount.

The Albigensian crusade initiated by Pope Innocent III in 1209 against what the church considered to be a heretical Christian sect in southern France was unprecedented in the atrocities it perpetrated. The pope's offer of indulgences to all who participated brought over twenty thousand Christians from all over Europe, all eager to stamp out the heresy.

This was the first time a pope had sanctioned a war against other Christians. At the siege of the city of Bezeirs an estimated 15,000-50,000 men, women, and children were slaughtered. The pope's appointee, Abbot Arnald Almaric was asked during the siege how he intended to separate the believers from the heretics in the city. "Kill all. God will know His own," was his response. Victims were tortured slowly before being burned. Many whose lives were spared had their eyes torn out. The war lasted more than twenty years and is estimated to have cost the lives of a million people.

The Christian Crusades against the Muslims to recover the holy land is estimated to have cost nine million lives, at least half of whom were Christians. The intolerance and utter lack of respect for human life exhibited in these "holy wars", all of which were approved and supported by the church, leads one to wonder how the message of love and forgiveness which is central to Christian doctrine could be so badly abused.

The French wars of religion which broke out in 1562 between Catholics and Calvinists (Huguenots) caused rivers to overflow with the blood of the dead. During the St. Bartholomew's Day Massacre Huguenots in Paris were tortured, shot, drowned, hanged and butchered by Catholics. Hunted down in their homes and workplaces, no one was safe. Incited by the Franciscan monks in the city, women and children were dragged through the streets naked before being killed and thrown into the Seine. The French Wars of religion caused an estimated seventy to a hundred thousand dead.

The cruelty of Spanish conquest and occupation of Amerindians in Central and South America is well documented. The Spanish soldiers tortured and killed mercilessly and brutally exploited the survivors as slaves in their mines. All was done with the encouragement and support of the church.

One must conclude that the doctrine of love and forgiveness in monotheistic religions takes second place behind the injunction of Jesus that "He who is not with me is against me."

One could argue that things have changed, that we have grown beyond the kind of religious acceptance of violence recited above but have we? Religious extremists today have been among the first to call for violence in the name of God, Jesus, or Allah. President George W. Bush stated publicly that before making the decision to attack Iraq he had consulted a "higher Father," clearly implying that God had approved the action which to date has resulted in the deaths of thousands of soldiers and tens of thousands of civilians. The religious conservatives who support him are among the most nationalistic and "hawkish" people in the country. One of their leading spokesmen, Pat Robertson, called for the U.S. to "take out" the president of Venezuela because he is "a terrific danger" bent on exporting Communism and Islamic extremism. Communism, because it is "godless" has been a favorite target of conservative Christians for decades. Muslim extremists regularly call for the killing of westerners in Allah's name and encourage suicide bombers. Nowhere in the teachings of Jesus or Mohammed is the call to violence justified, which demonstrates in dramatic form just how incapable organized religion is of living up to its own code of conduct. Robert Owen, founder of a utopian community at New Harmony, Indiana, had this to say regarding the depredations carried out in the name of religion:

James R. Curry

> *"All the religions of the world are based on total ignorance of all the fundamental laws of humanity. Fully conscious as I am of the misery which these religions have created in the human race, I would now, if I possessed 10,000 lives and could suffer a painful death for each, willingly thus sacrifice them to destroy this moloch."*

There will always be tension between those who see religion in terms of human growth and betterment and those who see it as a rigid, unchanging code of conduct to which people must conform, no matter the cost. This raging struggle goes on throughout the world, and it is a struggle which has the potential to tear the world apart. It is dangerous beyond belief.

> *There has never been a war fought over scientific principles.*

Scientists take it as a responsibility to disagree, and to settle disagreements through rational discourse. Out of these debates comes clarity and new perspectives. Putting someone to death because they do not accept a scientific point of view is antithetical to the culture of science, which is why I believe that in this respect at least, **science has the moral high ground over religion.** Considering the number of people who have suffered and died because of religious differences, I don't think the point can be challenged.

Exclusivity

Many religions contend that they and they alone, hold the keys to the kingdom of heaven, however envisioned. Anyone not practicing their religion is an infidel or non-believer. I reject this exclusivity out of hand. From the standpoint of the evolution of our species it makes no sense at all. Through the ages, perhaps reaching back beyond the birth of our own species, there have been tens of thousands of religions, all of which served for good or ill the people who practiced them. Any religion that brings moral guidance, dignity, and a sense of value and purpose to the daily lives of its practitioners can be beneficial, but to serve the needs of people a religion must be an integral part of culture. Forcing Christianity upon Australian aborigines (or other indigenous societies), for example, in no way helped them to interpret their world or survive in it. Their system of beliefs and practices were woven into their lives. Religious and spiritual beliefs were so integral a part of their everyday life that they dictated what people could eat, who they would marry, and the art work they painted or carved on implements and rock faces. They believed that the spirit of ancient ancestors resided in the land and in the people themselves, giving each person a strong sense of belonging to his tribe and the territory they claimed. It is thought that Aborigines have lived in Australia for at least 60,000 years and they numbered about a million people when the first white settlers arrived. They lived in small tribal groups, each with its own customs, beliefs, art forms, and ceremonies. By any standard, they were a highly successful culture, surviving and prospering in some of the harshest environments on Earth. Of what use would Christianity have been to such people? How could it possibly have contributed to their survival or well being?

Religious practices, symbols, and icons imaginatively detail what is important to a culture or subculture of people, but cultures do not remain static, and as they change religious practices and beliefs need to change with them. More than that, religion needs to instigate and lead, not follow cultural change. The failure of most modern religions is that they live in the past at the expense of the present and future. Religion, working through individuals, has an almost unrivaled power to change the direction of society. The profound and lasting influence of Abraham, Jesus, and Mohammed on society is obvious to every thinking person and does not need to be elaborated. Suffice it to say that the world would be a totally different place had these religious leaders never lived. In recent decades the social changes influenced by the activities of religious leaders like Mhatma Ghandi, Martin Luther King, Jr., and Desmond Tutu, while certainly not of the same magnitude, have helped to bring about significant social change which a majority of people would consider to be beneficial. Religion has the power to give people the strength to suffer adversity in order to right wrongs and free enslaved people.

Spirituality

The essence of religion is the will to believe. It is a free will choice. If we remove all the rituals, dogma, and icons that have accumulated over the millennia like barnacles on the keel of a ship, what is left but the pure, unadulterated will to believe? It is inherent in our species – a genetic predisposition as strong as our fear of the night or need for sociality. It is magnificent in its strength and purity. It is our will to believe which has sustained us throughout our history as a species.

God is beyond us, totally and fearfully beyond the scope of our senses.

We will never know what, if anything exists beyond the natural world. The ancients believed that trees, mountains, the sun, and other natural objects contained spirits. This is the origin of our spirituality, but no one has ever seen a spirit, and no one has ever returned from the grave to tell us if a spirit world exists. Skeptics and atheists point out this inability to confirm a spiritual world as a reason not to believe. I reject such reasoning. Our view of the world is limited by our senses. This is true of all animals. To a jellyfish there is no world beyond the ocean, and it is a world that is perceived only by touch and chemical sensors. A jellyfish does not marvel at the beauty of a sunset or revel in the graceful movements of a porpoise because it has no visual sense. Such things do not exist for a jellyfish. Every creature is aware of only those elements in its environment which have some meaning for its life. This combination differs for each species.

Animals of all kinds are aware of aspects of the natural world that humans cannot sense at all. Many pollinating insects have eyes that can see ultraviolet light. A flower that appears to us to be plain white is seen by them as purple with colorful markings. Bats send out ultrasonic vibrations which bounce off objects and are reflected back. This system of "sonar" is more sophisticated than any invented by humans. Thousands of bats swirling about in a completely dark cave are able to avoid collisions and fly safely out the entrance with the use of their sonar. Our senses do not permit us to hear these sounds. There is so much in the natural world that is beyond our senses: the infrared heat which rattlesnakes use to detect warm animals; pheromones used by many insects to attract mates; radio waves; atomic radiation; magnetism, and a host of other natural phenomena. Until instruments were devised which could convert these phenomena into sight or sound, we were unaware of their existence. One has to wonder what enigmatic wonders of our world still exist to be discovered. We are conscious of only the tip of the iceberg of information about the natural world.

We place entirely too much confidence in our ability to interpret the world through sense perception and mental imagery. The sky is blue only in our minds. Stimuli like light, sound, touch, etc. are converted by the various sense organs into an electrochemical signal which travels to the central nervous system. There, the information is processed and interpreted. We never "see" reality directly but only an internal representation of it (in our brain). The brain is like the captain of a submarine who is completely shut off from the outside world and must interpret the world outside the submarine indirectly from sonar and other sensing devices. The captain's interpretation of the information coming into him is largely determined by his training. Blips on a screen could be a whale, submarine, or mass of floating kelp. Likewise for each of us: we see what our culture and socialization predispose us to see.

We have always survived by imposing patterns on nature and propagating them in our children. These patterns, like the patterns on a submarine's sonar screen, allow us to navigate through the environment successfully.

Deep down I believe that we realize the inadequacy and artificiality of our mental constructs. What we perceive as reality is at best an abstraction. We are forever walled off from reality by the severe limitations of our sensory and central nervous systems. In the words of the poet Wallace Stevens, "*We live in the description of a place and not in the place itself.*" Biologists know this better than anyone. To categorically deny that there are elements in nature that we are not able to confirm with our senses is not credible scientifically.

In the same sense that a tree exists within me and is defined by the design of my sensory organs and central nervous system, so perhaps does God reside in me, defined and limited severely by my humanity. Imagine a fish swimming near the surface of the water when an airplane flies over. To the fish the plane is merely a dark moving spot which is blurred and obscured by the water, and in the brain of the fish there is no comprehension of what it is seeing. The size, shape, and sound of the plane mean nothing to it. Still, it is aware of something, no matter how muddled and incoherent it may be. Perhaps we are like the fish when it comes to perceiving the supernatural. Perhaps what is supernatural to us is merely an extension of the natural world which we perceive in the same way the fish perceives and comprehends the airplane.

For now we see through a glass, darkly; but then face to face: now I know in part; but then shall I know even also as I am known.
1 Corinthians 13: 13

Faith is personal - a personal perception of something only very dimly perceived by our senses. It is an intuition which can grow and develop, becoming deeper and richer with the years. There is no real way to explain faith except to say that it fulfills a human need to believe that there is something greater than one's self. Religion is a social way of expressing faith. Religions are built upon cultural traditions which have been passed down for thousands of generations and which serve the needs, however poorly, for people to express their faith.

Faith gets confused with religion and religion gets confused with church temple, or synagogue. Being highly socialized animals, our faith has naturally sought an outlet in a social setting and churches, synagogues, mosques and temples have served that purpose. Unfortunately, these religious social settings have all the failings of any socio-politico-economic contrivance. Infighting is common, cliques form and compete with each other for control, and the constant requests for money and time can cause people to become frustrated and confused about the real purpose of their membership. Often people walk away feeling, "I don't need this: there are already too many complications in my life. I come here to get away from all of that." Hopping and shopping for an amenable religious venue is a common phenomenon among people seeking a place of refuge from everyday problems.

On the Other Side of the Curtain

"Why are we squandering the talents of women, fully half of all God's creation?"
     - Irshad Manji

The major religions have become so encumbered by tradition and ritual that they make slaves of their practitioners and strangle their faith. Every religion has a canon of beliefs to which the body must subscribe. Often these beliefs are as outmoded as a slingshot in modern warfare, but tradition demands that they be adhered to. The role of women in organized religious life is one of these. Christianity, Judaism, and Islam have traditionally been resistant to making women equal partners in leadership positions. In so doing, they have kept women subservient to men and squandered the creativity and productivity of fifty percent of the human family. Today, moderate elements in these three great religions are slowly moving toward granting women greater leadership roles, but the fundamentalist and ultra-conservative religious institutions continue to be especially sexist.

In Judaism women are separated from men by a curtain or in a balcony during prayer and they are forbidden to participate in many aspects of the service. Until very recently, a woman could not transcribe a Torah. One of the ancient prayers still said by men thanks G-D for "not making me a woman." Women generally have an inferior place in those synagogues, as the following quote by a Jewish woman indicates:

> "While I don't personally feel the pull to lead davening, get an aliyah or become a rabbi, I understand the frustration at such a closed door. My issues focus more on how disproportionately smaller the women's side of the Kotel is, or how men's yeshivas get so much more funding than women's yeshivas, and how noticeably more joyous the reaction from people is when a baby boy is born, etc. I comfort myself with the hope that these are merely anachronistic remnants of an older era, and not religious creed. Nonetheless I sometimes can't help but feel that with the admittedly important exception of eventually making babies, Judaism doesn't really need me. At least not as a single person. I'm not needed to complete a

minyan, it is not incumbent upon me to pray three times a day, nor will I ever be asked to be a witness at the wedding of a close friend."
(Laya: www.jewlicious.com/index/php/conference) 4-15-05

"Laya" goes on to state, *"Yet in so many ways Jewish women were given more rights and respect than women in almost any other culture."*

The position of women in Islam in theory differs vastly today from Islam in practice. Mohammed said, *"All people are equal like the teeth of a comb. There is no merit of an Arab over a non-Arab or a white over a black person or of a male over a female."* Another of his teachings is, *"Women are the twin halves of men."* Statements granting equality to men and women in education and the everyday aspects of life abound in the Koran. Unfortunately, the Koran has many contradictory and ambiguous passages. Selective interpretation by patriarchal leaders and the mingling of Islamic teachings with tribal customs and traditions have perverted what was once an enlightened role for Islamic women.

Although the Koran speaks of men and women as being spiritual and intellectual equals, women must enter the mosque by a different door than men and the sexes are separated during worship. Women are not permitted to lead prayer lest their comely looks distract men from their prayers. Forcing women to cover their bodies from head to toe with just the eyes exposed is also a means of preventing them from arousing men. Women who are raped in Muslim countries are sometimes put to death as "adulteresses." One has to wonder why Muslim men are so fearful of the allure of women.

Should a religion be judged by its teachings, or by the actions of people who claim to be its followers? In Afghanistan under the Taliban, women were not permitted to attend school, practice a profession, be seen with any part of the body other than the eyes exposed, drive a vehicle, or leave the house without the permission of their husbands.

In Saudi Arabia, the religious police stopped schoolgirls from leaving a burning building because they were not wearing the headscarves and abayas required by the kingdom's strict interpretation of Islam. It was reported that three policemen were observed beating young girls to prevent them from leaving the school because they were not wearing the abaya. Fifteen girls were burned to death as a result.

Islamic extremists today seem bent on self destruction. The declaration of fataws on anyone who speaks ill of Islamic teachings or practices, suicide bombings against Israel and the west, and the ascendancy of extremism all speak of a religion and society that questions itself to the point of being suicidal. Unless moderate and liberal Muslims seize the lead and press for moderation and modernization in Islam, the self-hatred which permeates it today will eventually draw the entire world into a violent bloodbath unprecedented in the history of the world.

> *Sir, a woman preaching is like a dog's walking on his hind legs. It is not done well; but you are surprised to find it done at all.*
> - Samuel Johnson.

Early in the history of Christianity, women played major roles. They were among Jesus' most devoted followers. When he was arrested and his male disciples fled, women followed him to the cross. It was two women who discovered his empty tomb and first saw the Risen Jesus. During the first centuries of the early church, many more women than men were converted. Some of these women played prominent leadership roles, as documented in the writings of Paul, and were an important factor in its early success.

Despite the importance of women at the birth of Christianity, they have since been marginalized and continue to be in the fundamentalist and ultra-conservative denominations today. Church leaders defend this policy of discrimination with Biblical Scripture like these:

> *"Let the woman learn in silence with all subjection. But I suffer not a woman to teach, nor to usurp authority over the man, but to be in silence."*
> - I Timothy 2:12-13

> *"Let your women keep silence in the churches: for it is not permitted unto them to speak; but they are commanded to be under obedience also saith the law. And if they will learn anything, let them ask their husbands at home: for it is a shame for women to speak in church."*
> - I Corinthians 14: 34-35.

Faith should bring people together: beliefs often prevent that from happening. There is no better example of this than the inability of organized religions to fully integrate women into church structure and worship. The Biblical injunctions above were written at a time when a woman's role in society was limited to bearing children and caring for her family under the domination of her husband. Parents prayed for boy children and mourned the birth of girls, believing that girl babies were a punishment for their sins. Women had no legal rights, could not own property, and were little more than slaves to men. To let the writers of two millennia ago dictate the role of modern women in the life of the church is as great a folly as it would be to allow doctors to use the medical treatments of that day on their patients.

And yet, that is what is happening as men stubbornly hold onto the reigns of ecumenical power.

On February 27, 2006, at the opening of the 50th session of the United Nations Commission on the Status of Women, Secretary General Louise Frechette said that empowering women and girls around the globe is the most effective tool for a country's development. She went on point out that studies have repeatedly shown that giving women equal educational and work opportunities and participation in decision-making processes can increase a countries economic productivity, reduce infant mortality, and improve general nutrition and health.

Gender equality is absolutely essential in achieving the United Nation's Millennium Development Goals (MDG's), a blueprint for improving the lot of the world's poorest people. All 191 members of the UN have pledged their support for meeting the eight goals by 2015. They are as follows:

Eradicate extreme poverty and hunger.

Achieve universal primary education.

Promote gender equality and empower women.

Reduce child mortality.

Combat HIV/AIDS, malaria and other diseases.

Ensure environmental sustainability.

Develop a global partnership for development.

None of these goals are even remotely possible without releasing women from the suffocating yoke of male domination which exists throughout the world today. Women make up fully seventy percent of the 1.3 billion people who live on less than $1.00 per day. They also make up 2/3 of the world's illiterate, and own only 0.01% of the world's property.

> *I have this sense that Jesus tried to give us something, but his gift was stolen by the proud and power-hungry.*
> \- Tom Ehrich

In March, 2006, Pope Benedict XVI, in response to a question by a young priest, stated that the Vatican will consider increasing women's "institutional" role in the church, but that they would remain barred from the priesthood. He argued that priests manifest the teachings of Christ, who chose men as his disciples. The Roman Catholic Church remains one of the most male dominated of all the Christian churches, which helps to account for the fact that Catholic dominated countries in South America, the Philippines, and elsewhere remain among the poorest on Earth, have some of the highest rates of illiteracy, infant mortality, and population growth, and remain in abject poverty despite being rich in resources. The Vatican's unwavering stance against abortion and family planning keeps women in perpetual bondage to their husbands and children and causes population growth rates to soar out of control. The extremely authoritarian nature of church doctrine stifles creative thinking or individual action and keeps church dominated societies in perpetual stasis. The opulence of the cathedrals, wealth and power of the church, all stand in stark contrast to the poverty and helplessness of the people.

In 2004, Pope John Paul II called on leaders of the Roman Catholic Church to attack feminist ideologies which assert that men and women are fundamentally the same. He expressed the concern that the belief is eroding women's maternal vocation. Antiquated attitudes and beliefs like these are destructive to the UN'S Millennium Development Goals and to the hopes and prayers of billions of poor people around the globe. The Catholic Church is rapidly becoming powerful in Africa, whose poor countries lead the world in poverty, HIV/AIDS, illiteracy, infant mortality, population growth rates, violence toward women, and all the other degradations associated with gender inequality. The church's policies toward women can only exacerbate these problems in the coming decades. The needs of people are obviously secondary to the goals and ambitions of the church.

Blessed are the poor: for your's is the kingdom of God.
- St. Luke 6: 20

To understand how organized Christians devoted to improving the lot of people can perpetuate such destructive beliefs, one has to look at them not as religious institutions, but as political and economic institutions seeking power, money, and influence. They do this by being active politically, and by being the most aggressive religion on Earth at evangelizing. There are no countries where Christian evangelists are not actively seeking to increase the membership and power of their particular church. I am convinced that policies like those against female empowerment are really directed at increasing membership and power by more or less forcing women to be birth machines. Study after study has shown that the most effective way to reduce abortion, increase literacy, improve public health, and decrease poverty generally is through the empowerment of women. Churches which are totally male dominated at all levels simply refuse to share their power with women, thus reducing them in third world countries to a life of child bearing and poverty.

Jesus and the founders of other great religions did not take their gospel to the wealthy and powerful. They appealed to the downtrodden, the despised, and the powerless. The gospel of Jesus empowered his outcast followers to the point that they were eventually able to "conquer" the mighty Roman Empire. But, as Christianity conquered Rome, Rome converted Christianity, and the Church took on the legal, political, and economic trappings of the Roman Empire, thus inaugurating that quest for money, power, and influence which prevails today.

> *Most men are harassed and buffeted by life, and crave supernatural assistance when natural forces fail them; they gratefully accept faiths that give dignity and hope to their existence, and order and meaning to the world.*
> - Will Durant

Are we fools? Are we fools to believe that there is a world beyond our senses, a world we can only dimly perceive? Have our fathers back to the beginning of time been totally fooled into believing in the spirit world? If so, why do even non-believers so often turn to God in times of crisis, people like the sailors aboard the aircraft carrier Franklin when it was turned into a raging inferno by Japanese fighter bombers? A group of men were trapped below decks in total darkness and thick smoke as massive explosions ripped the ship apart from one end to the other. The hatch they tried to escape through was locked down and so hot they couldn't touch it. All thought they would die, and they began to recite the Lord's Prayer. None of them knew it all, but between them they completed it. Eventually sailors above deck heard their cries for help and opened the hatch, freeing them.

Later in the war when the USS Indianapolis went down, the surviving men floated for four days before being rescued. Sharks attacked them constantly, killing many of the survivors, while many others died of thirst. Out of 900 men who went into the water only 316 survived. One of the survivors said "There were a lot of men who didn't know God when they went into the water, and there were a lot of men who wanted to know God once they were there."

James R. Curry

THE BOONDOCKS © 2004 Aaron McGruder. Dist. by UNIVERSAL PRESS SYNDICATE. Reprinted with permission. All rights reserved.

In 2003 a young hiker named Aron Ralston was trapped in a remote Utah canyon when a heavy rock rolled onto his arm. Alone, and with almost no food or water, he survived for five days before deciding to cut off his arm with a cheap knife which had been dulled by his attempts to chip away at the rock. It took about an hour to cut through the arm below the elbow. Once released, he crawled through a narrow, winding canyon, rappelled down a 60-foot cliff, and walked six miles before meeting other hikers and getting help. In a meeting with reporters he later described his experience, during which he frequently made references to prayer and spirituality. "I may never fully understand the spiritual aspects of what I experienced, but I will try," he said. "The source of the power I felt was the thoughts and prayers of many people, most of whom I will never know."

There are thousands of stories of people turning to spirituality when caught in a seemingly hopeless situation. Is turning to God in times of crisis a weakness or strength? If we lose our faith what is left in times of crisis but despair? If belief in the supernatural is simply a matter of choice, why do so many people world-wide make that choice? I believe it is because spirituality gives us the ability to transcend ourselves.

There is much talk these days about a "God Gene." No one really believes that a single gene could be responsible for our interest in the supernatural, and yet the universality of this interest bespeaks a genetic predisposition. Is our belief in the supernatural, as Stephen J. Gould and others profess, simply a byproduct of other capabilities associated with our sociality? Or, as theologians maintain, is it because we are made in the likeness of God? There is nothing about our species that is completely unique; not our bodies, not our behavior, not our brains. The differences that exist between us and other creatures are matters of degree, not uniqueness. Could it be that we are not alone in our appreciation of the supernatural?

Once, when in a hurry, I opened the kitchen door without remembering that our cat George often waited nearby for the chance to prowl the outside world. Like a flash he was out the door. It was late March, and a brilliant male cardinal was standing on the patio outside the door. The cat hit the bird with a speed I didn't know he possessed. I chased him off as quickly as I could, but it was too late: his teeth had penetrated the body cavity and the cardinal lay on his side, crest raised, red blood spilling out onto red feathers. I felt responsible for the bird's wounds and tenderly picked him up. If I had not been negligent in letting the cat out this beautiful little animal would still be able to sing his song, care for his mate, and raise his young. Not wanting to let the cat get him again, I laid him on the branches of a cedar tree out of reach and went on my way. When I returned home about a half hour later I looked at the cardinal, expecting it to be dead. It wasn't moving, but I could see that it was still breathing. At that moment, the bird rolled onto its feet and began to beat its wings violently, its beak open and crest raised in defiance of death. I knew it was in its death throes, and my sense was that the bird knew it was dying and was fighting against it. The fight lasted only a few moments and death came quickly. Up until that time I had never wondered whether an animal could be sufficiently aware of its own existence to have a sense of its approaching death. I still think to this day that the cardinal knew it was about to die.

Since then I have wondered about the ability of other living things to have a self awareness. I wonder, for example, how a tree "feels" when it is being consumed in a forest fire, or how an earthworm feels when it is drying up on a sidewalk. I also wonder if other living things are more responsive to the supernatural than we are. Humans are denied the knowledge of much that goes on in the natural world because of the limitations of our senses. Many animals (and plants?) sense aspects of nature that we are completely unaware of. Did we lose some of our receptivity to supernatural events when we lost our tail and fur? Perhaps our intuitions about the supernatural are mere vestiges of a once powerful life force which still infuses the lives of other living things. Religion has torn us away from our relationship with the natural world. Our animistic ancestors felt a oneness with the Earth and its life forms which we now consider to be primitive and pagan, but which gave them a sense of belonging that we lack. We are strangers in our own world, interlopers searching for a place that does not exist in this life.

Attendance is declining in many parts of the world today in organized religions, but spirituality is not. This frightens religious leaders who see in this trend a loss of power, money, and interconnectedness. Religion is a social activity, but spirituality is very personal. A quiet revolution is in the making, one which has the potential to bring back the sense of mystery, excitement, and fear which the ancients felt in their world of spirits. People may turn their backs on organized religion, but the imperative to recognize the supernatural is too deeply implanted in our natures to ever disappear.

> *Indeed, I do not know whether, if our reverence for the gods were lost, we should not also see the end of good faith, of human brotherhood, and even of justice itself.*
> - Cicero

CHAPTER 24

Spreading the Gospel

Religious Predators

Just one more channel. One more channel to spread the gospel of Christ. Help us by sending a donation today. Pick up the phone right now, and God will bless you.

If you have no money, if you are thousands of dollars in debt, send us a hundred dollars or a thousand. If you do so, if you make this commitment to the gospel of Jesus Christ, God will wipe out your debts and make you prosper.

God is speaking to you. He is telling you to get out of the hundred flow and get into the thousand flow.

We have an army of prayer warriors around the world. Send us your prayer requests with a donation.

Give what you can't afford and God will return it to you many fold.

Do you think God would have any trouble getting $1,000 to you somehow?

When you give to God you're simply loaning to the Lord and he gives it right on back.

The people making these pleas ride in limousines, fly in private jets, wear clothing that the average person would work months to pay for, live in mansions, and stride about opulent television sets wearing outlandish hairdo's and makeup. And yet, the contributions pour in, underwriting a lifestyle for the televangelists that most of the people who support them will never enjoy.

I am bewildered by all this. It is patently obvious to even the most casual observer that the televangelists are not promoting the kingdom of God; they are building their own private kingdoms – monuments to themselves. Deep down, I suspect the televangelists know this, but their egos are such that they always want more: more money, more power, more public exposure, or one more channel for God.

Not all televangelists are in this mold. Some are undoubtedly people of integrity who passionately believe that what they are doing is helping people find their way to the kingdom of God. Even so, televangelists are forever asking for money, and the people who give the largest gifts (as a percent of their income) are the people who can least afford it. Why then do so many people support them? What do their supporters get in return that is so important to them, and why aren't there more legitimate ways for people to satisfy their needs?

The ease with which televangelists raise millions of dollars causes many people to support them in the belief that their prosperity is evidence of God's blessing. Other supporters believe they are helping to save billions of souls by making it possible to send the gospel of Christ to the world. And for many supporters, the programming offered is a lifeline that is unavailable to them elsewhere. Praying with a minister on television may give momentary relief for whatever problems a person is suffering, but person to person interaction would be so much more effective and lasting. What is lacking in our society that forces people to get support from television and to send money they can't afford to religious predators?

"Speed Bump" © Dave Covery/Dist. By Creators Syndicate, Inc.

The Mayans Had It Right

In contrast to the wealthy televangelists are those dedicated people, both ministers and laity, who give up their American lifestyle to live in a small village in Guatemala, Africa, or elsewhere in order to spread the gospel. They are passionate about what they do, and they are willing to live without running water, indoor toilet facilities, heating and air conditioning, and all the other comforts of life that Americans take for granted. Often they open small infirmaries in villages that have no medical services, provide food, clothing, books, and other necessities supplied by their support group in the U.S. By comparison with the wealthy televangelists they are to be respected for their sacrifice and passionate beliefs.

But, why do they do it? The question nags at me. Why the extraordinary need to change other people's religious beliefs and practices? Why are they willing to belittle what others believe, and attempt to change them to our beliefs? Why is Christianity, in contrast to most of the other major religions, so competitive and aggressive in its efforts to evangelize?

Historically, Christian missionaries have destroyed indigenous cultures and religion at the point of a gun. Religion is an integral part of the culture and environment of every society. The beliefs, rights, and practices are often thousands of years old in development, and they reflect the life style and social, economic, and political practices of the society. Destroying a society's religion tears apart the entire fabric of their culture.

Christian missionaries know this, if only subconsciously, for they work to change culture to suit Christian teachings. Robert Fitzroy is an example of why this approach doesn't work. His efforts to introduce Christianity to the natives of Tierra del Fuego in the 19th century were as absurd in that indigenous culture as Jemmy Button's top coat and hat.

Go ye therefore, and teach all nations, baptizing them in the name of the Father, and of the Son, and of the Holy Ghost.
- St. Matthew 28: 19

Christ's admonition to his disciples following his resurrection became the rallying cry of a missionary movement that quickly spread the gospel into the entire known world. Beginning with Paul, the movement was so successful that within three centuries, and against all odds, the entire Roman Empire had become Christian. The extraordinary success of the early missionaries confirmed the rightness of their cause in the minds of Christians and led to an ever increasing effort to convert the world; an effort which continues to this day.

The protestant foreign missionary movement in America began early in the 19th century with the formation of the American Board of Commissioners for Foreign Missions in 1810. Prior to this time most missionary efforts were focused on converting the native Indians, but this had little success. In the decades which followed missionaries were sent to India, the Sandwich Islands, Africa and other far distant lands where souls waited to be saved from the fiery pit of hell. The movement grew to fever pitch as passionate evangelists began returning with hair-raising stories of their efforts to bring Christ to the pagans. Congregations across the country raised money and sent missionaries into foreign lands. The efforts increased dramatically following the Civil war and have not abated since. The passion and aggressiveness of Christian missionaries stems from the belief that no other faith can save people. Only those who accept Christ as their savior can expect to reign with him in Heaven. What is more, some in every generation since Christ promised to return have believed that he would do so during their time. This lends a sense of urgency to the effort to convert "lost" souls before it is too late. Today, evangelists of fundamentalist and ultra-conservative faiths are proclaiming that all of the signs pointing to the return of the Savior are coming to pass in our generation, and those who do not heed the warning signs and give their lives to Christ will be forever lost.

The Christian belief that we are born sinners, without redemption save we be baptized in the name of Christ, is a curious one. From a biological or psychological standpoint this is nonsense. We are born innocent of all evil, but with a need for training and discipline if we are to develop into socially functional and useful people. It makes no sense to think that honest, caring, well adjusted people of any culture will suffer eternal damnation because they were born in sin and are not members of a particular religion. Even as a child I could not accept that I was born a sinner and would die a sinner unless I accepted Jesus Christ as my savior. To my mind, religious leaders have propagated this self-serving belief for the same reason that Mayan or Egyptian leaders propagated the belief that they were gods: to gain and maintain control.

> *From Greenland's icy mountains, from India's coral strand;*
> *Where Afric's sunny fountains roll down their golden sand:*
> *From many an ancient river, from many a palmy plain,*
> *They call us to deliver their land from error's chain.*
>
> *What though the spicy breezes blow soft o'er Ceylon's isle;*
> *Though every prospect pleases, and only man is vile?*

Children of God: Children of Earth

In vain with lavish kindness the gifts of God are strown;
The heathen in his blindness bows down to wood and stone.

Shall we, whose souls are lighted with wisdom from on high,
Shall we to those benighted the lamp of life deny?
Salvation! O salvation! The joyful sound proclaim,
Till earth's remotest nation has learned Messiah's Name.

Waft, waft, ye winds, His story, and you, ye waters, roll
Till, like a sea of glory, it spreads from pole to pole:
Till o'er our ransomed nature the Lamb for sinners slain,
Redeemer, King, Creator, in bliss returns to reign.

"Mission Hymn". Reginald Heber, 1819.

The result of missionary efforts in foreign lands has been the destruction of traditional religions and cultures and the acceptance of Christianity and western culture. The natives of the Sandwich Islands, now Hawaii, became early targets for conversion by American missionaries. With the support of The American Board of Commissioners for Foreign Missions a company of New Englanders made up of two ministers, a doctor, two teachers, a printer, their wives and five children, sailed from Boston on October 23, 1819. They reached the Islands at a time of internal strife which made the natives receptive to the newcomers and their religion. By 1860 dozens of churches and schools had been established, and the initial goal of Christianizing Hawaii had been achieved. The successful efforts of this small, committed group of people to destroy the indigenous culture are still evident in Hawaii today.

Not only have we failed to bring them the news of Christ with sincerity and honest faith, but we have betrayed in our deeds what we professed in our words.
        - Jose de Acosta

Acosta was a Jesuit missionary in the Caribbean and South America during the late 16th century. During his travels he kept a journal in which he described the horrific treatment inflicted upon the indigenous people by the Spanish. He particularly objected to missionaries condoning such actions, and to using the Indians as slave laborers.

Missionaries have often been at the forefront of empire building. This was true of the Spanish missionaries in South and Central America and the Philippines. It was also true of the English missionaries in India and Africa. Chinua Achebe, in his widely acclaimed book, *Things Fall Apart,* describes the destruction of the native Ibo religion and culture in Nigeria at the turn of the

20th century. When the first English missionaries arrived they were ignored or treated with scorn. When they asked for land to build a church they were offered part of the "evil forest", where was buried anyone who had died of a terrible disease and which was a dumping ground for the potent fetishes of great medicine men. The natives avoided the forest believing that to enter it meant death. To their surprise, the missionaries accepted the land and began building a church. The Ibo people expected them to be dead in 4 days, and when they were not, they thought they had unbelievable power.

The missionaries went among the people saying that they must forsake their false gods of wood and stone and serve the one true god or they would burn in hell. "If we leave our gods and follow your god…who will protect us from the anger of our neglected ancestors," the natives asked. Once the church was built evangelists were sent out to surrounding towns and villages. The clan leaders believed the white man's god would not last. The early converts were men of low esteem in the village, "worthless, empty men." The priestess called the converts the "excrement of the clan," and the new faith was a mad dog that had "come to eat it up."

The whites built a school and hospital, then set up a court system and began to imprison the natives for traditional actions like putting twin infants in the forest to die. The judges and others did not speak the native language and made no effort to understand traditional land rights and practices. The first missionary was conciliatory, but he was replaced with one that was uncompromising in his zeal to destroy the "pagan" practices of the people. One of the native Christian converts tore off the mask of a "spirit" during a native ceremony celebrating the earth deity. It was an act on a level with murder, and in response the natives destroyed the church. "It was a terrible night. It seemed as if the very soul of the tribe wept for a great evil was coming – its own death." The tribal leaders were tricked into meeting with the white authorities and arrested. Their heads were shaved and they were shackled and mistreated. In the end, the natives had no choice but to give in to the new religion and the white man's governance.

Christian missionaries historically only rarely stopped to consider the value of other religions or of the culture they represent. Consider, for example, the story of creation from the *Popol Vuh*, the sacred book of the Maya Indians of Central and South America. In the beginning there was only the sky. First the gods created the earth and covered it with trees and other plants, but they were dissatisfied that all was in silence and without movement. To fill the void they created the four-footed animals and birds and assigned to each a specific habitat and niche within the forest. They commanded them to speak and glorify the gods, but all the animals could do was to scream and make noise. Disappointed, the gods condemned them to a life of cruelty: they would be hunted, torn apart, and eaten.

The gods, still desiring to be praised and adored, decided to make men. First they made him of mud, but his flesh was soft and he could not move. He could speak but had no mind. They destroyed him. Next the gods made man of wood. The wooden men soon covered the earth with their kind, but they had no souls or minds, and they walked on all fours. Worst of all, they could not know their creator. They destroyed the wooden men.

In their third and last attempt they made man of corn mixed with the blood of the gods themselves. The men were intelligent, handsome, and able to comprehend the world as well as the gods. They were far seeing and glorified the gods for creating them and giving them the power to see all and understand all.

But the gods were not pleased that the men were their equals and they destroyed their wisdom and knowledge. They then created wives for them so that they could multiply and populate the earth. The following selected passages from the Popul Vuh are reprinted with permission of the University of Oklahoma Press.

This is the account of how all was in suspense, all calm, in silence; all motionless, still, and the expanse of the sky was empty.

This is the first account, the first narrative. There was neither man, nor animal, birds, fishes, crabs, trees, stones, caves, ravines, grasses, nor forests; there was only the sky.

The surface of the earth had not appeared. There was only the calm sea and the great expanse of the sky.

There was nothing brought together, nothing which could make a noise, nor anything which might move, or tremble, or could make noise in the sky.

There was nothing standing; only the calm water, the placid sea, alone and tranquil. Nothing existed.

There was only immobility and silence in the darkness, in the night.

Then while they (the gods) meditated it became clear to them that when dawn would break, man must appear. Then they planned the creation, and the growth of trees and the thickets and the birth of life and the creation of man.

Thus let it be done! Let the emptiness be filled! Let the water recede and make a void, let the earth appear and become solid; let it be done. Thus they spoke. Let there be light, let there be dawn in the sky and on the earth! There shall be neither glory nor grandeur in our creation and formation until human being is made, man is formed. So they spoke.

Then they made the small wild animals, the guardians of the woods, the spirits of the mountains, the deer, the birds, pumas, jaguars, serpents, snakes, vipers, guardians of the thickets.

And the creation of all the four-footed animals and the birds being finished, they were told by the Creator and the Maker and the Forefather: "Speak But they could not make them speak like men; they only hissed and screamed and cackled; they were unable to make words, and each screamed in a different way.

When the Creator and the Maker saw that it was impossible for them to talk to each other they said: "It is impossible to say our names, the names of us, their Creators and Makers."

Children of God: Children of Earth

Then they said to them: "Because it has not been possible for you to talk, you shall be changed. Your food, your pasture, your homes, and your nests you shall have; they shall be the ravines and the woods, because it has not been possible for you to adore us, we shall make other (beings) who shall be obedient. Accept your destiny: your flesh shall be torn to pieces. So shall it be.

They wished to give them another trial; they wished to make another attempt; they wished to make (all living things) adore them.

But they could not understand each other's speech; they could succeed in nothing, and could do nothing. For this reason they were sacrificed, and the animals which were on earth were condemned to be killed and eaten. For this reason another attempt had to be made to create and make men by the Creator, the Maker, and the Forefathers.

"Let us try again! Let us make him who shall nourish and sustain us! What shall we do to be invoked, in order to be remembered on earth? We have already tried with our first creations, our first creatures; but we could not make obedient, respectful beings who will nourish and sustain us."

Then was the creation and the formation. Of earth, of mud, they made [man's] flesh. But they saw that it was not good. It melted away, it was soft, did not move, had no strength, it fell down, it was limp, it could not move its head, its face fell to one side, its sight was blurred, it could not look behind. At first it spoke, but had no mind. Quickly it soaked in the water and could not stand. And the Creator and the Maker said: "Let us try again because our creatures will not be able to walk nor multiply. Let us consider this," they said. Then they broke up and destroyed their work and their creation. And they said:"What shall we do to perfect it, in order that our worshipers, our invokers, will be successful?"

They said, "Say if it is well that the wood be got together and that it be carved by the Creator and the Maker, and if this [man of wood] is he who must nourish and sustain us when there is light when it is day! "So may it be," they answered when they spoke. And instantly the figures were made of wood. They looked like men, talked like men, and populated the surface of the earth.

They existed and multiplied; they had daughters, they had sons, these wooden figures; but they did not have souls, nor minds, they did not remember their Creator, their Maker; they walked on all fours, aimlessly. They no longer remembered the Heart of Heaven and therefore they fell out of favor. It was merely a trial, an attempt at man. At first they spoke, but their face was without expression; their feet and hands had no strength; they had no blood, nor substance, nor moisture, nor flesh; their cheeks were dry, their feet and hands were dry, and their flesh was yellow. Therefore, they no longer thought of their Creator nor their Maker, nor of those who made them and cared for them.

These were the first men who existed in great numbers on the face of the earth.

Immediately the wooden figures were annihilated, destroyed, broken up, and killed. A flood was brought about by the Heart of Heaven; a great flood was formed which fell on the heads of the wooden creatures

So was the ruin of the men who had been created and formed, the men made to be destroyed and annihilated; the mouths and faces of all of them were mangled. And it is said that their descendants are the monkeys which now live in the forests; these are all that remain of them because their flesh was made only of wood by the Creator and the Maker. And therefore the monkey looks like man, and is an example of a generation of men which were created and made but were only wooden figures.

After that they began to talk about the creation and the making of our first mother and father; of yellow corn and of white corn they made their flesh; of corn meal dough they made the arms and the legs of man. Only dough of corn meal went into the flesh of our first fathers, the four men, who were created.

They were endowed with intelligence; they saw and instantly they could see far, they succeeded in seeing, they succeeded in knowing all that there is in the world. When they looked, instantly they saw all around them, and they contemplated in turn the arch of heaven and the round face of the earth.

James R. Curry

The things hidden [in the distance] they saw all, without first having to move; at once they saw the world, and so, too, from where they were, they saw it.

Great was their wisdom; their sight reached to the forests, the rocks, the lakes, the seas, the mountains, and the valleys. In truth, they were admirable men…

They were able to know all, and they examined the four corners, the four points of the arch of the sky and the round face of the earth. But the Creator and the Maker did not hear this with pleasure. "It is not well what our creatures, our works say; they know all, the large and the small," they said "What shall we do with them now? Let their sight reach only to that which is near; let them see only a little of the face of the earth! It is not well what they say. Perchance, are they not by nature simple creatures of our making? Must they also be gods? And if they do not reproduce and multiply when it will dawn, when the sun rises? And what if they do not multiply?" So they spoke. "Let us check a little their desires, because it is not well what we see. Must they perchance be the equals of ourselves, their Makers, who can see afar, who know all and see all?" Thus they spoke, and immediately they changed the nature of their works, of their creatures. Then the Heart of Heaven blew mist into their eyes, which clouded their sight as when a mirror is breathed upon. Their eyes were covered and they could see only what was close, only that was clear to them.

In this way the wisdom and all the knowledge of the four men, the origin and beginning [of the Quiché race], were destroyed.

In this way were created and formed our grandfathers, our fathers, ...

Then their wives had being, and their women were made. God himself made them carefully. And so, during sleep, they came, truly beautiful, their women, ..

> From Popul Vuh: The Sacred Book of the Ancient Quiche Maya. English version by Delia Goetz and Sylvanus G. Morley from translation of Adrian Recinos. Copyright © by the University of Oklahoma Press

The Maya peoples dominated Mexico and Central America for a thousand years. Originally hunter-gatherers from the north, they moved south into the lush tropical forests and prospered. Four thousand years ago they made the transition to agriculture, and with this more reliable and productive method of getting food their numbers began to grow. Corn was the basis of their entire culture. As their staple crop corn enabled them to populate and conquer Mexico and all of Central America. Corn enabled them to develop a complex society with astronomers, mathematicians, artisans, and priests. They produced a calendar that was accurate to within one day every 3600 years. At their peak they numbered at least 10 million people. The wealth of their empire was dependent on corn and the riches of the forest. This was reflected in their religion. The Jaguar was a sacred animal and his figure permeated their culture and religion. The great Ceiba tree of the forest connected the heavens to the underworld: the spirits of the dead entered its roots to descend from the underworld to heaven above.

The Maya owed their existence to maize and the blood of their gods. Their most important gods were the god of rain, the sun god, and the god of corn. They worshipped these gods because they were the gods that made them successful by giving them corn and the wealth of the forest.

In hundreds of cities and towns the Maya raised magnificent temples of stone in which to worship the gods and seek their pleasure. Their religion was tied intimately to their environment and to their way of life. Christianity would have had absolutely no meaning for them because it did not spring from the soil of their culture and being. It would have been as alien and meaningless to them as the pagan rites of hunter-gatherers are to us.

The tropical soils are among the poorest on earth for agriculture. Once the forests are cut down the heavy rains quickly wash away the soil down to a rock hard subsoil. Once this happens, soils which once supported great tropical forests are able to support nothing more than desert plants. As the Maya overpopulated the land more and more of the forests were converted to farmland and areas of human habitation. In order to build their magnificent temples, grand plazas, and houses they made mortar out of crushed and burnt limestone, burning up tens of millions of trees in the process. By 1,100 years ago most of the land was cleared. The forest, upon which they relied to supplement the products of their agriculture, was gone. The impoverished soil led to the collapse of their agriculture, and starvation ensued.

As resources declined wars were fought over what remained, and human sacrifices were used in supplication to the gods. The ruling elite were unwilling to relinquish their lifestyle and power, and they made no voluntary compromises which might have slowed the decline into poverty. War, famine, greed, and abuse of the land all lead to the demise of the once wealthy and flourishing Maya empire. The jungle eventually reclaimed their great cities to such an extent that their very existence was lost to the world for centuries.

Almost all of the written records of the Maya were burned by the Spanish as "heretical works," a loss to the world of incalculable value. The *Popul Vuh* was saved because some Maya priests and clerks secretly made copies of it in Latin (Use of the Maya script was forbidden by the Spanish and Latin was taught instead), and hid them in the mountains. One of these was discovered by a Spanish priest who translated it into Spanish.

The Maya religion explained natural forces that were not well understood, paid homage to what was most important in their life and society, and justified the authority of the ruling elites by making them representatives of the gods. Religion was an integral part of the fabric of their culture, as it is of the cultures of all indigenous peoples. Forcing Christianity or any other religion on such people makes no sense in terms of their religious and spiritual needs. Most of the Maya people alive today worship in the Catholic religion, but it is a blend of Christian and indigenous beliefs. Some of the churches and cathedrals which the Spanish built were constructed on sites holy to the Maya in an effort to get converts. To this day the people remember the location of these sites and perform Maya rituals on those precise spots within the church building.

North America needs more missionaries. Right now we have only one North American missionary for every 43000 lost people. OnMission.com

The arrogance and exclusivity of Christianity as expressed by western cultures historically have been the downfall of many indigenous cultures around the world. Some, like the American Indians, have resisted the strong arm call to Christ, but most have fallen before its force. How tragic that Christ's commandment to bring to the world God's love, mercy, and justice has led instead to the destruction of the beauty, creativity, and functionality of indigenous religions.

CHAPTER 25

Demons

When I was a child I came to know a man named Ishmael. Ishmael was a strange old man, but he was a friend of my parents and they valued his friendship. From time to time he visited us, usually in the evening, and he and my father would talk long into the night. Ishmael was tall, with the build of a scarecrow. He was so thin that his Adams apple seemed to project too far out from the rest of his body. He had a full head of thick gray hair which he kept cropped short, and a gray mustache. When he walked he seemed to lean over from the waist so that there appeared to be a hump in his back.

Ishmael always appeared to be a rather sad man. I don't remember ever seeing him smile or laugh. His health was not good and he never worked at a job as long as I knew him, nor did he do chores around the house. Mostly what Ishmael did was to sit and brood about things most people would rather not think about. He was not always like that. As a youth he was energetic and imaginative. At the age of twelve he built a parachute out of bed sheets and jumped off a railroad bridge down onto a paved road nearly a hundred feet below. Everyone who heard of it said the same thing: "It's a wonder he wasn't killed." He wasn't, of course, or I would never have known him. Apparently his parachute, if it can be called that, was sufficiently well designed to break his fall and he landed without injury.

Later, as a teenager, Ishmael designed and built an airplane out of wood and wire. He did this without any research, relying on his own native intelligence and concept of what an airplane should be. He had no engine for the plane, so it never got off the ground, but word got around and professors from Carnegie Tech (now Carnegie-Mellon University) came and inspected the plane and were apparently quite impressed because they offered him a full scholarship on the spot. Unfortunately for Ishmael, and perhaps for the

Children of God: Children of Earth

world, his father was a very religious person, and he believed that a university degree would turn Ishmael from the Lord, so he was not permitted to accept the scholarship. We can only guess what Ishmael's life might have been had he gone to Carnegie Tech and gotten an engineering degree, but as it was Ishmael never really amounted to much by the standards of our society. He did, however, raise a large family, all of whom turned out to be very fine, productive people. This was due more to his saintly wife, who bore the poverty and endless work without complaint.

As I mentioned, Ishmael never worked, and this gave him more time to himself than perhaps a person should have. Like all of us, Ishmael fought demons, but unlike most of us, in the long brooding hours, days, weeks, and years, his demons became real. They came to him in the dark of night, and he would wrestle until dawn, leaving Ishmael too exhausted to do more than sit on the porch all day.

One night while he and my father were talking, Ishmael began to tell Dad about wrestling with the demons. We lived in a very small house, all on one floor, and conversations in one room were easily heard in another. My brother and I and a cousin my age were in bed, and as the two men talked we listened to Ishmael's stories of his bouts with demons.

Being seven or eight years old at the time, and not very knowledgeable about such things, we listened with growing terror to the stories, and it was long after Ishmael left that night before we fell asleep. For a long time after that we slept with sticks and an old pocket knife beside the bed in case the demons came for us. I am glad to report that they never did pay us a visit, but to this day I can remember the terror I felt in the darkness of that bedroom wondering when I might see a dark shape moving silently through the room and have to fight for my eternal soul.

Ishmael lived to a ripe old age, so I guess his fights were generally successful. I believe that he fought demons all his life - we all do - but I am hard put to explain how he came to believe that he actually fought with demons from hell.

The truth is, the only demons are the ones that exist in our own minds. Satan is a figment of the imagination of ancient people who are long lost in the dusty mists of time. It seems to be a quirk of human nature to need to see things in opposites – life is defined by death; good is defined by evil; change is defined by permanence, happiness by sadness, health by illness, and so forth. The opposites present each other in sharp relief; each alone would be amorphous and relatively without meaning.

So, fear not demons, but do not ever give up the fight with your personal demons, for to do so is to give up on life. I think Ishmael found his personal demons too difficult to face, and so he invented demons he could fight with, demons so powerful that losing to them was no reason for guilt or shame. Those who knew him as a young man were saddened that his obvious potential for worldly success and self-fulfillment were denied, and they overlooked his shortcomings. He was, in fact, liked by everyone who knew him. He was a good man by all accounts.

In thinking about demons I am reminded of a story about the great French naturalist, Georges Cuvier (1769-1832). Working at the Paris National Museum of Natural History, Cuvier firmly established the fact of extinction of past life forms, single-handedly founded vertebrate paleontology as a scientific discipline, and created the comparative method of study of organismal biology. He did not, however, believe in the evolution of living things. He studied mummified cats and birds brought back from Napoleon's invasion of Egypt and showed that they were in no way different from cats living at the time. He used this to support his belief that organisms were functional wholes, and that any change in one part would destroy their ability to survive; hence, evolution through time was impossible.

> *Why has anyone not seen that fossils alone gave birth to a theory about the formation of the earth, that without them, no one would have ever dreamed that there were successive epochs in the formation of the globe.*
> - Cuvier, "Discourse on the Revolutions of the Surface of the Globe."

Cuvier had an uncanny ability to reconstruct organisms from fragmentary fossil remains. Although some earlier scientists believed that fossils represented life forms now extinct, the prevailing belief of the day was that God would not permit life forms which He had created to be destroyed. Cuvier believed that the Earth was immensely old, and that periodic catastrophes (he called them revolutions) had occurred, each of which wiped out a large number of species. He did not believe that new species arose to take the place of those lost to extinction.

But, back to demons. Cuvier's students once dressed up as demons and crept into his bedroom in the middle of the night. Standing beside his bed they began a mournful chant: "Cuvier, Cuvier, we are going to eat you." Awakening, Cuvier looked at them through sleepy eyes and said matter-of-factly, "Animals with horns and hoofs are all vegetarians. You can't eat me," then turned over and fell back asleep. So, there you have it from one of the world's great scientists: demons are vegetarians; they cannot possibly harm you!

The Devil is a Lonely Man

I had a sister, and her hair
Was pale as hair could be:
Her eyes were green as caves half-seen
Beneath a quiet sea.

The devil is a lonely man;
My sister had a whim.
The devil was a lonely man,
And she could comfort him.

She built a bonfire on the beach,
And waiting, said a prayer.
(The sea at highest tide could reach
And kiss her footprints there.)

But she could conjure in the night
As only children can,
And so she called him to her side.
(He was a lonely man.)

He watched with her for seven hours
And told her many lies-
Or were they truths – the awful things
That make her now so wise?

She often sits beside the sea.
The vulgar would discover
(But cannot, since she never speaks)
How brutal was her lover.

She sits and stares into the flames
And silently she cries,
But all her tears can never veil
The doom before her eyes.

For she has learned his loneliness,
Nor ever needs be told
How one can burn one's face with fire
And still one's back is cold.

James R. Curry

Come up, my little sister,
Come up from by the shore.
But no, she does not come nor sing
Nor conjure any more.

She shares the devil's loneliness –
If ever woman can.
But he, despite her sharing it,
Is still a lonely man.

                        Lois Anne Davison

CHAPTER 26

Do Unto Others

"It is only in the giving of ourselves to others that we truly live."

Ethel Andrus

The great religions of today were borne of the agricultural revolution. The introduction of agriculture spelled the doom for most hunting-gathering societies, along with the rules of behavior which had been so important to the survival of human groups from the beginning of time. Most hunting-gathering groups were small, often only several dozen people, most of who were bound together by a familial relationship. With agriculture came cities, land ownership, commerce, social hierarchies, politics, and laws. The unwritten rules of social behavior which had served humanity so well in small mobile societies failed in the dynamic heat of large urban societies where people had to compete with strangers for survival, and where accumulating and protecting goods was more important than interdependence and sharing.

The story which follows is reprinted with the permission of W.O. Pruitt, Jr. and Lyons Press/Globe Pequot Press. It describes the kind of selfless cooperation and altruism which hunting-gathering societies relied upon for their survival. The Dinje, or People of the Moose, were Athapaskan Indians who lived in the tundra and great northern forests of Canada and Alaska. Their Winter Festival, or *Tcitciun*, celebrated their commitment to each other in a land where no one person can survive alone. Each hunter had a hunting partner (kla) who was his constant companion. He shared everything with his kla and would willingly risk his life for him.

"The people now abandoned their moss houses and gathered into a village. Here each family made its first winter house – a hemispherical framework of poles covered with furry moose hide, and heaped around outside with an insulating layer of snow. The floor was of spruce twigs, redolent and tangy. In the center of each house was a hearth and above the hearth was the smokehole in the roof. The air within the winter house remained clean and fresh because the great temperature differential between inside and outside caused fresh air to be constantly drawn in through the snow.

The Moose People began the festival of *Tcitciun* with a round of storytelling. All during the twenty-hour nights they would sit entranced, adults and children alike, while an old man would recount the people's history of the great days when the god-man Tsa-o-sha lived among them, of the hunter who wrestled and killed the grizzly bear and brought its meat to the hungry people."

Each night the people gathered in a different house and each night the children learned more of their heritage. They learned why the grizzly was respected, why the otter was never killed, and why the shrew was to be feared above all else. They learned to forecast the weather by the loon's wail and to interpret the raven's shout when a moose was in his view. The stories and dicta were the strands of culture and heritage that bound the people into a complete entity. After a complete recital of manners, and morals of the Dinje, the people devoted two days to *Tcitciun* or the Hook Game. The hunter and his family sat in their house. Suddenly the caribou skin door was thrust aside and a long stick poked in. Suspended from the stick was a length of babiche onto which was tied a wooden hook. The hunter shouted a man's name – no response. Another name – no response. Finally he shouted the correct name and the hook jiggled. Next he shouted the name of an object – dried moose meat – no response. Flaked flint adz? No response. Basket of lingenberries? The hook jiggled. The basket was put on the hook and it disappeared out the door.

All through the day the hunter and his family, laughing uproariously, gave whatever the hooks wanted, until by nightfall they had virtually no possessions left. Next day was their clan's turn to visit and dangle the hook. When they finally returned home they again possessed a complete array of tools, clothing, and food, some new, some old, but all *gifts*. Thus, just as the gift custom bound the hunter to his kla with bonds of gratitude, so *Tcitciun* bound all Moose people together. No matter how far apart they might travel, each Dinje knew that he was a brother to all other Dinje with whom he had played the Hook Game. The ceremony of the Winter Festival was his peoples' way of acknowledging that no person can stand alone. By exchanging gifts each member of the small community was bonded to all the others in their

Children of God: Children of Earth

struggles to survive in this land of evergreen trees, tall mountains, and barren tundra." Pruitt, W.O. Jr. 1988. *Wild Harmony: Animals of the North.* New York: Lyons Press/ Globe Pequot Press. Copyright 1967.

The Athapaskan Indians of Alaska and northern Canada did not judge each other by how much they owned, but by how much they were willing to share with each other. They had few material goods, and these needed to be constantly repaired or acquired anew for they were a people constantly on the move as they hunted, fished, and gathered food with the changing seasons. Like all hunter-gatherers, they lived in small groups, had no political or social hierarchy, lacked designated leaders, and based their social relationships on cooperation and mutual trust and respect. They had no written laws or other rules of conduct by which to regulate behavior, and yet their society functioned smoothly, and for the most part, was without crime or warfare. The Winter Festival which reinforced their ties to each other epitomizes the cooperative behavior which is the essence of our humanity.

It was a hard life because of the severity of the climate, but the Athapaskan Indians thrived on it and prospered. Their pattern of life was based upon cooperation, mutual respect, and interdependence. No one ever went hungry while others had food, no one attempted to accumulate more than his neighbors or to wield power over them. No one stole from his neighbors or refused to share with them. No one thought of the land or the plants and animals on it as his own. The life style of the Athapaskan Indians was, in most respects, typical of hunter-gathering societies in general.

The agricultural revolution was, without any doubt, one of the most far-reaching and rapid transformations ever undergone by any species of living organism. The change in survival strategy from hunting-gathering to the domestication of animals and planting and harvesting of crops apparently occurred within a few thousand years and it shifted much of the burden of adapting to a changing environment from biological to cultural evolution. Cultural evolution has proven to be extraordinarily rapid and progressive. In less than fifteen thousand years our species has grown from a world-wide population of about five million to almost seven billion, and we occupy and dominate all areas of the globe.

Biological evolution took tens of millions of years to produce our species. Change was excruciatingly slow; often millions of years passed with no visible change in our pre-human ancestors. Cultural evolution, spurred on by the riches of agriculture and vitality of technology, took us from the plains of Africa and caves of the Middle East and Europe to global domination in 10,000 years. People who had lived in small, scattered, nomadic tribes less than five thousand years earlier were building magnificent structures like the

Great Pyramid of Cheops and living a lifestyle radically different from their predecessors.

The preadaptations which made such a colossal change in the direction of evolution possible were in the development of our hands and brain. Hands which could throw a spear could also build the Great Pyramid. A brain which could learn to control and utilize fire and invent a spear thrower could visualize such a complicated and enormous building as a pyramid and could develop the organization required to transport materials to the work site, feed tens of thousands of workers daily, and provide governmental supervision of the entire project.

As human societies grew in size and complexity it became necessary to codify and enforce rules of conduct which stone-age peoples took for granted. The code of Hammurabi in Babylon is an early example of efforts to control antisocial behavior through written laws (Hammurabi ruled Babylon from 1792 – 1750 BCE). The growth of civilization brought with it crime, violence, and warfare unlike anything experienced by hunter-gatherers as nation states vied for land and resources and people competed for wealth and power. Killing, pillaging, and plundering replaced the cooperative, self-sacrificing nature of earlier societies. Clearly the social rules had changed dramatically and it seemed as if the species had turned down a dead-end road to destruction.

Agriculture and the birth of civilization brought to the fore the worst elements of human nature and subdued the best. People who had lived in small, nomadic tribes, who owned no land and very few possessions, and who more often than not were all related in some way, had little difficulty maintaining harmony in their societies. Their dependence on each other was clearly understood, and they shared everything they had. This was how they survived. Agriculture and civilization enabled people to stop wandering and settle in cities and towns. This in turn allowed them to accumulate food and other possessions, and to use these as sources of wealth and power. Competition replaced cooperation, and selfishness replaced altruism.

Karen Armstrong in her book *The Great Transformation* points out that during the period from 900 to 200 BCE the great religious and philosophical beliefs –Confucianism,Buddhism, monotheism, Hinduism and philosophical rationalism – were created which have continued to inspire people through the ages. The German philosopher Karl Jaspers coined the term Axial Age for this period of time because of its importance to the future development of philosophy and religion. Armstrong goes on to indicate that although Christianity, Islam, and Judaism later developed from these original intuitions, the world has never grown beyond them.

What the great sages of the Axial Age recognized was that human society was in a period of intensely rapid transition which required that religion, which up until that time had involved primarily animal sacrifice and animism, now had to step forward to bring order out of the social chaos wrought by the development of civilization. Armstrong states that "During this period of intense creativity, spiritual and philosophical geniuses pioneered an entirely new kind of human experience. Many of them worked anonymously, but others became luminaries who can still fill us with emotion because they show us what a human being should be." With this I disagree. The Axial sages and those who followed them did not create something totally new; they were reinventing and reinforcing human standards of behavior as old as humanity itself. Consider the following quotes, some of which are from the writings of the great prophets or their followers who came later:

> *The man of perfect virtue, wishing to be established himself, seeks also to establish others; wishing to be enlarged himself, he seeks also to enlarge others.*
> Confucianism. Analects 6.28.2

> *All men are responsible for one another.* Judaism. Talmud, Sanhedrin 27b

> *The best of men are those who are useful to others.*
> Islam. Hadith of Bukhari

> *Let no one seek his own good, but the good of his neighbor.*
> Christianity. 1 Corinthians 10:24

> *The sage does not accumulate for himself.*
> *The more he uses for others, the more he has himself.*
> *The more he gives to others, the more he possesses of his own.*
> *The way of Heaven is to benefit others and not to injure.*
> *The way of the sage is to act but not to compete.*
> Daoism. Tao Te Ching 81

> *The ignorant work for their own profit, Arjuna: The wise work for the welfare of the world without thought for themselves.*
> Hinduism. Bhagavad Gita 3.10-26

> *Without selfless service are no objectives fulfilled; In service lies the purest action.*
> Sikhism. Adi Granth, Maru, M.1, p.992

James R. Curry

He who prays for his fellowman, while he himself has the same need, will be answered first.
Judaism. Talmud, Baba Kamma 92a

Do nothing from selfishness or conceit, but in humility count others better than yourselves. Let each of you look not only to his own interests, but also to the interests of others.
Christianity. Philippians 2. 3-4

If you will now go back and reread the story of the Winter Festival at the beginning of this chapter, you will discover that the Athapaskan Indians lived by the rules of conduct set out in the quotes from these various religious writings. Indeed, these are the same rules by which pre-civilization societies lived or they would not have survived. The only thing new in them is that the Axial sages and their secondary followers in later centuries codified them in writing and reaffirmed their importance to the individual and to society. More importantly, perhaps, they expanded religion to include the regulation of social behavior. Animism related the individual to the natural and supernatural worlds but not to his fellow man. The sages recognized that unless the peoples of their day returned to the *behaviors* of the past, the increasing violence of their world would destroy humanity. They chose religion as a way of achieving this critical goal.

How well has the goal been achieved? One need only look about at today's world to realize that the violence and selfishness by individuals and nations is as great today as in the day of the Axial sages and great prophets. What worked well for the hunting-gathering peoples has not translated to civilized societies. Some of the very strengths which helped the ancients to survive: strong group identity, cooperation within the group and distrust of other groups, are tearing us apart today. The Axial sages saw the need to rise above identification with a clan or tribe to identification with all of humanity. The concept that all men are brothers which permeates the beliefs of all the major religions has never been achieved in any sense by civilized people. Nationalism, regionalism, religious affiliation, racial identification; all are born of the tribal mentality and all thrive today to keep us apart and at war with each other. We answer the question "Am I my brother's keeper" with a resounding, no.

Civilization replaced the classless society of hunter-gatherers with a rigid structure of elites who ruled and peons who served. Empires replaced clans and armies replaced hunting partners. Conquest and warfare replaced relative peace, and respect for the natural world was replaced with ownership and exploitation. The Axial sages were attempting to stem the violence of their society and bring it back to the rules of conduct which had made humanity successful for hundreds of thousands of years. Unfortunately, their idealism has failed overwhelmingly. The record of humankind in the past ten thousand years is one of continual warfare, social strife, and environmental degradation.

Are we by nature capable of loving our neighbor as ourselves? I believe we are, as evidenced by people like those of the Athapanska. The reason why the major religions have failed is because they have set ideals which are unworkable in civilized human society. We no longer live in small, related groups. Most of us spend our days with unrelated people who are competitors for jobs, salary increases, recognition, and friends. We are taught from childhood to be individualistic and competitive. Success is not measured in terms of human relationships but in terms of wealth and power. Ownership and status are more important than cooperation and brotherly love. This is the nature of civilization; this is what is dictated by land ownership and the political and economic systems which developed from agriculture. As Daniel Quinn points out in his acclaimed book *My Ishmael*, we are all captives of a civilizational system that more or less compels us to act as we do. He likens us to tigers in a cage that pace back and forth relentlessly and cannot understand why they are bound by the bars which restrain them.

In places… where animals are penned up, they are almost always more thoughtful than their cousins in the wild. This is because even the dimmest of them cannot help but sense that something is very wrong with this style of living. The tiger you see madly pacing its cage is nevertheless preoccupied with something that a human would recognize as a thought. And this thought is a question: Why? "Why, why, why, why, why, why?" The tiger asks itself hour after hour, day after day, year after year, as it treads its endless path behind the bars of its cage. It cannot analyze the question or elaborate on it. If you were somehow able to ask the creature, "Why what?" It would be unable to answer you. Nevertheless, this question burns like an unquenchable flame in its mind, inflicting a searing pain that does not diminish until the creature lapses into a final lethargy that zookeepers recognize as an irreversible rejection of life.
MY ISHMAEL: Bantam edition, Random House, Inc.

The majority of people in the world long for something better than civilization has brought us. We are all tigers in a cage, isolated, frightened, confused by the seeming irrelevance of our lives and our inability to find our way back to our true selves. The constant yearning for rapture, heaven, or nirvana is an irreversible rejection of life brought about by an overwhelming sense of helplessness, but it is more than that; it is a statement of how we would prefer to live: with compassion, cooperation, love, and peace.

Hunter-gatherers lived in small groups with a high degree of relatedness. We, on the other hand, live in a world in which the majority of the people we deal with are non-relatives. Apart from family and a few friends to whom we would trust our lives we treat other people in a way that serves our interests and goals. We respect and admire the strong and revile the weak. We choose leaders whose alpha male nature leads us into conflict when negotiation, diplomacy, and respect for others would serve us better. Except for a miniscule number of people in history – people like Christ, Mohammed, or more recently people of the stripe of Gandhi and Mother Teresa - we all protect what is ours and strive to manipulate others for our benefit. The people of the world today are like a giant dysfunctional family.

A Failure to Adapt

The problems and challenges which beset civilized man have never changed. Technology changes, but greed, egotism, hatred, sectionalism, racism, war, crime, and hunger are unresolved after 10,000 years of civilization. They are systemic to our civilization. The lesson of the Athapaskan Indians and other hunter-gathering peoples is that cooperation, respect for others, and altruism are a part of our nature which we apply freely to relatives and to those who reciprocate in kind. Most hunting-gathering groups numbered between 30-50 people, many or most of who were related to each other in one way or another. The behaviors which enabled groups like these to survive evolved over millions of years, and our species would not have survived without them. Civilization took us in a radically new direction, one for which our behavior was only minimally adapted. Behaviors which worked well in small closely related groups have not prepared us to relate well to each other in cities of thousands or millions of people, all but a few of whom are strangers, competitors, and potential threats in one way or another. The less structured life of hunter-gathering groups gave way to the incessant toil of farming, and sharing gave way to protecting what is mine: my land, my animals, my tools, my food. As commercialism grew, so did greed and the desire to accumulate and control money and goods. The leaderless, classless nature of ancient peoples gave way to highly layered societies which were largely controlled by the wealthy and powerful. The history of the last 10,000 years is one of masses of people being driven to nation building and conquest by small numbers of highly aggressive people who have taken over leadership roles and used their positions to attempt to achieve more wealth and power.

The paradox is that overwhelmingly people prefer cooperation to competition; prefer peace to warfare; prefer to be their brother's keeper to being his enemy; prefer honesty to guile and manipulation; prefer altruism to greed; prefer compassion to cold indifference. These positive behaviors come to the fore during periods of crisis when people forget their own needs and concerns and focus on helping others. Why is it not possible to act this way toward others all of the time? Part of the reason, as Armstrong points out, is that we are "conditioned to self-defense. Even within our own communities and families, other people oppose our interests and damage our self-esteem, so we are perpetually poised – verbally, mentally, and physically – for counter attack and preemptive strike." This is as true of nations as it is of individuals. All too often our good deeds are met with suspicion because we all protect ourselves and strive to manipulate others for our benefit. It has often been pointed out that nations do not have friends, only interests. To some extent that is also true of people.

Thou shalt not kill lest it be to the sound of martial music.
- Bertrand Russell

It is in our nature to be protective and cooperative with people we identify with and suspicious and potentially hostile toward those who are "outsiders." Group identification is a powerful behavioral modifier. During World War II "Japs" were depicted in the U.S. as subhuman creatures who needed to be destroyed. Today the Japanese people are among our closest allies and friends because under western influence during the occupation of their country they became enough like us to become part of our identity group. Group identity often relies on demonizing those not in the group. This makes it easy, even patriotic or desirable, to treat them inhumanely. The fire bombings of Japanese and German cities which killed hundreds of thousands of people in the most inhumane manner possible are cases in point. The men in the American and British bombers were not killers in the ordinary sense. They were men who were protecting their "group" against the threats from an "out group". If this artificial distinction between us and them could have broken down these airmen could not have committed such a tragedy. The American people turned against the Vietnam War largely because of a now famous photograph of a young girl running down a road with her clothes burned off from napalm dropped by air force bombers. The terror and pain on the youngster's face was something people everywhere could understand and identify with and they were reviled by it.

Rituals promote group cohesion by requiring members to engage in behavior that is too costly to fake.
- Richard Sosis

All religions promote in-group cohesion and identity by encouraging or requiring their members to adhere to behavior which identifies them with their particular religious group. Through the use of special rituals, clothing, ornamentation, greetings, facial and/or head hair, dietary requirements, prayers, and other means members demonstrate their commitment to their religious group. This kind of behavior builds trust and cooperation within the religious group and sets them apart from other groups. To further show their commitment, the members of religious groups often are hostile to other groups, castigating them as heathens, nonbelievers, infidels, pagans, or other such derogatory terms. At its worst inter-group hostility among religions has lead to "holy wars" such as the Crusades, the Inquisition, and the Spanish wars on the indigenous peoples of Central and South America. Hostility between religious groups continues to this day to be a serious problem locally and globally.

Group thinking must be replaced with personal relationships if we are ever to achieve the kind of a world everyone so desperately wants. We need to recognize and emphasize our humanity, not our special interest groups. We need to feel that we are one people whether we are or not. We need a stated goal – one we cannot possibly live up to today, but one which will serve as a beacon toward which we can move. To my mind some of the greatest words ever written were by Thomas Jefferson in the Declaration of Independence: "We hold these truths to be self evident, that all men are created equal, that they are endowed by their Creator with certain inalienable rights, that among these are life, liberty, and the pursuit of happiness." This was written by a man who held slaves all of his life. Could there be a greater hypocrisy? And yet these words have guided our actions in the centuries since they were written, and although we have not completely achieved this lofty goal, we have made significant progress toward it. We need a similar statement and sense of purpose to guide us in our efforts to achieve greater understanding and social harmony among all the peoples of the world.

Many of the problems between religious groups today stem from the need to reinforce in-group identity. The knee-jerk reaction of fundamentalist Christian groups to abortion, their persistent efforts to remove the teaching of evolution from the public schools, their overly loud objection to the removal of the Ten Commandments from public property, and their protestations of the removal of prayer from schools and public meetings are all designed to reinforce in-group identity while vilifying other groups of different faiths or beliefs. Many, perhaps most, fundamentalist Christians have only a glimmer of understanding about biological theories of evolution but they steadfastly deny their validity in order to demonstrate loyalty to their religious group. The outrageous protests of Muslims to any depiction of the prophet Mohammed, or to any criticism of their faith, is likewise a demonstration of commitment which reinforces in-group bonds at the expense of out-group relationships. The exaggerated nature of all these demonstrations clearly reveals their true purpose in promoting group cohesion.

"Speed Bump" © Dave Covery/Dist. By Creators Syndicate

The world community needs to commit to a course of action that will lead us toward a more perfect harmony, one not based upon Do Unto Others in the traditional sense, but one based upon a common identity. I am not suggesting that religious groups need to give up their special rituals and behaviors in order to achieve social harmony and cooperation. I am suggesting that we need to stop emphasizing our differences and emphasize the elements common to all religions. Instead of trying to live up to the impossible goal of loving others as ourselves we must make it our goal to understand others. This is a more powerful incentive, one which forces us to take the time to see the other's problems and point of view. With understanding will come compassion and a willingness to join with people of different faiths in a common effort to promote cooperation and social harmony on a global scale.

Original Sin?

Behold I was shapen in iniquity; and in sin did my mother conceive me.
- Psalm 51:5

The concept of original sin is a non-starter. We are not evil by nature or birthright. To believe that because Adam and Eve sinned the entire human family is condemned to hell unless we as individuals redeem ourselves is to allow a fable older than time to enslave us with guilt and fear. The whole concept of sin is an absolutist way of thinking that does more harm than good. The Axial sages and prophets who followed them were obsessed with sin. We would be far better off to redefine sin in relative terms of survival. Trying to hold ourselves to an absolutist definition of what is moral and what is sinful makes hypocrites of us all: murder is a sin; but murder in defense of country is patriotic and blessed of God. Lying is a sin, but telling a lie to avoid hurting someone's feelings, or lying to protect the national interest is morally correct. To some people being gay is a sin: to others, making a likeness of the prophet Mohamed or a woman not wearing a head covering is a sin. The world needs to move beyond this slavish adherence to bigotry and narrow-minded thinking and begin to emphasize the powerful underlying themes of love, forgiveness, humility, and brotherhood which are the keystone of the great religions. The concept of sin is often destructive of social harmony because it sets people against each other. Battles over abortion, gay sex, germ cell research, and a host of other social issues are the result of absolutist thinking regarding sin.

We need to forget sin and focus on developing children who will grow up to be caring, respectful, thoughtful individuals, people whose lives will be guided not by a rigid set of rules, but by a mature ability to relate to other people in a way that promotes harmony and mutual understanding. We need to relate to other people based upon how they behave, not how they think or what they believe. A bigot in mind who treats all people with sympathy and respect is no sinner. A man who admires beautiful women but never cheats on his wife or disrespects women in any way is not a sinner. We cannot really control what we think, but we can and must control how we behave. Trying to live up to injunctions like "love others as yourself", or "don't look upon a woman with lust because you will have sinned in your heart" is a lesson in futility and guilt.

The focus of religion needs to change from the individual to society, from sin to interpersonal relationships, from the afterlife to this life, because unless our species can move to a higher plane of sociality, one in which communication, cooperation, and mutual respect replace tribalism and sectarian violence, the future will be one of increasing social disintegration on a global scale. Nothing less than our survival as a species is at stake.

I am convinced that if a revolution in our social structure is to take place it will have to be a grass roots movement. Relationships between governments are amoral and impersonal. Lying, cheating, manipulating, killing; all are acceptable in the name of national interest, and all are routinely condoned and practiced by government officials who think of themselves as upright moral people. I am impressed by how many times comments are made by people in countries attacked by the U.S. to the effect that they don't blame the American people, they only blame our government or our President. People everywhere have the same hopes for themselves and their children, weep over the same tragedies, and share the same joys. Gestures that reach out to people mean so much more than government to government interactions. Ping pong in China and soccer in Iran resonated with the common people of those countries in a way that no amount of foreign aid by our government could have accomplished. Small towns across this country have ties with "sister cities" in Japan, Europe, and other areas of the world. Why not join hands with towns and cities in parts of the world hostile to us? Relationships like these can be extremely beneficial in bringing people together in a common understanding. Religious groups can play a leading role in bringing the peoples of the world together but only if they abandon their proselytizing in favor of unconditional altruism and service.

Seven billion people cannot live on this small planet thinking only in terms of self, community, or country. It is destroying us. We must find a way to unite into a spiritual whole. This does not mean giving up regional or social distinctness. It means reducing friction at the edges so that a unified quilt-work of relationships develops. It means rejecting exclusivity and refocusing away from technology and consumerism to human relationships on a global scale. The most critical aspect of any animal social group is communication. The higher the degree of sociality within a group, the greater the need for rapid, honest, unambiguous communication among its members. Communication is the cement which binds social groups together. The recent revolution in our communication systems: satellites, cell phones, e-mail, and the world-wide web, present us with the opportunity to break down the barriers which have always separated us from each other. Governments can no longer control information the way they have in the past. People all over the world are now in touch with each other and can have a person to person relationship even though they are on different sides of the globe or political spectrum. This bodes well for a movement away from confrontation and force and toward mutual cooperation, but only if we choose to make it so.

I am convinced that there are strong genetic undercurrents in our Stone Age nature which if allowed to surface and prosper can lead us into a more advanced social and political structure, one which will begin to move us in the direction of universal peace. This is unlikely to ever occur unless all of the peoples of the world rise up in protest and say "enough".

Altruism – Selfish Cooperation?

The Athapaskan Indians were able to live cooperatively because each member of the group understood the necessity to do so. In a sense, their behavior toward each other was self-serving. Unselfish behavior is common among animals which live in social groups, but it is difficult to explain in terms of natural selection. Vampire bats which have fed well will share their blood meal with another member of their group which has failed to find a meal. Worker bees will sting an intruder at the nest, thereby giving up their own life for the lives of others. Ground squirrels will sound an alarm at the sight of a predator even though doing so means drawing attention and perhaps the attack upon itself. Why would any animal engage in behavior which benefits others at their own expense? Natural selection, after all, favors genes which selfishly promote the survival and reproduction of the individual, not of others. These questions bedeviled Darwin because he recognized that altruistic behavior could undermine his theory.

Two theories have emerged to explain how cooperative genes can survive and spread in an environment of selfish genes. Vampire bats feeding each other demonstrate both of these theories. The first is that self-sacrificing behavior can bring a reward in the form of reciprocation. In other words, the well fed bat will at times be unable to find food and can expect to be fed by those it has fed in return. Studies have shown that for reciprocal altruism to work, the benefit to the receiver must exceed the cost to the donor, and donors must be able to recognize cheaters (for example, bats which accept blood but refuse to give it). A bat with a stomach full of blood loses little by giving away some of it, but could well be saving the life of the bat which receives the gift. The same bat which was generous on one occasion might itself be saved from starvation at a later date by another member of the colony. This sort of cooperation benefits all members of the colony. The second theory is encapsulated by a statement the British geneticist J.B.S. Haldane is said to have made while drinking with friends:

I would not risk my life for a brother but I would do so for two brothers or eight cousins.

No doubt Haldane was either joking or exaggerating to make a point. The point is that Haldane and his brothers and cousins share many of the same genes. Brothers have 50% of their genes in common. By saving two brothers Haldane would ensure that his own genes would pass into the next generation, the criterion for biological success. Each of his cousins, on the other hand, shares only 12.5 % of his genes; thus it would take eight of them to ensure that his genes made it to the next generation. Mutant genes which predispose an individual toward altruistic behavior can survive and spread through a population of social animals when that behavior is directed toward relatives who also have copies of the gene.

Social behavior has evolved in a way that promotes altruistic behavior toward close relatives or toward non-relatives which may reciprocate altruistic behavior.

Altruistic behavior as a result of kinship relationships is demonstrated among ground squirrels. These small animals live in groups in the western United States. Individuals within the group often assume the role of sentry by climbing a small tree or rock and watching for predators. If a hawk, eagle, or mammalian predator like a coyote is sighted the sentry gives a sharp whistle which cause a flurry of activity as all members of the group scurry for the tunnel. In giving a warning the sentry draws attention to itself and is at greater risk of being killed by the predator. In fact, sentries are killed more often than others in ground squirrel groups. Most sentries are females: males only rarely give an alarm call. Studies have shown that young males leave their group and join other groups while females generally remain with their group. This means that females have many relatives in the group while males have none. The willingness of a squirrel to put itself at risk is directly proportional to how closely related it is to the other members of its group.

Does any of this have relevance for altruism in humans? Sometime when you have nothing better to do, sit down and make a list of the people you would be willing to donate a kidney to. I don't think the list will be very long and the majority of the names on it will be close relatives. The only other names on the list will probably be people who, for one reason or another, you depend upon and would not want to lose; people in other words with whom you have a reciprocal cooperative relationship. This raises a question: is self-sacrificing behavior as a result of kinship or reciprocation truly altruistic? Or does all such behavior have an underlying basis in selfishness in the sense that the risks taken or help offered also confer a hidden benefit?

This is not to say that unconditional altruism does not exist. Clearly it does. There are many stories of people who have risked their lives to work with the sick or to save a stranger from death. Such behavior does not fit the pattern of reciprocal or kin altruism. There is obviously much that biologists don't understand about the evolution of altruistic behavior. Perhaps when we have unraveled this enigma we will better understand humanity's failure to adapt well to the changed social conditions brought about by agriculture.

Religious doctrine asks us to deny our human nature. We have not survived by denying our nature but by emphasizing it. We don't have to transcend ourselves to behave well. There is nothing wrong with acting cooperatively or compassionately towards others with the hidden, and perhaps unconscious, goal of somehow benefiting by it. Each member of an Anthapanscan Indian group was willing to cooperate with the other members of the group because he knew he could not survive alone. For selfish reasons we must learn to share, to understand, to compromise, to sacrifice – to do unto others for selfish reasons; not for a reward in Heaven, but for an immediate reward or the future promise of a reward that will somehow benefit me. Asking people to "do good" for a reward in heaven is hollow, and for the most part it doesn't work.

Jesus injunction to a wealthy man to give all that he had to the poor and follow him was rejected by the man. The man simply could not give up his hard earned wealth for something as intangible as a reward in Heaven. And yet many people do give huge sums of money for charitable purposes. Ask any development officer of a charitable institution why people are willing to do so, and they will tell you that the donor gets a tangible reward for the gift. The reward may be a name on some structure, the good will of people whose admiration he values, the publicity which lets the world know that he has wealth enough to share, or simply a "good feeling". Unconditional altruism in philanthropy is as rare as it is in other arenas of human activity. Selfish altruism because of kinship or reciprocity is the basis for most altruistic acts.

Ministers can talk about the need to transcend ourselves by rising above greed, egotism, hatred, and violence but it has little meaning for people in their everyday lives. Even well meaning people find themselves fighting for what is theirs in our competitive social structure. We are truly captives of a culture in which individualism and material success are valued above cooperation, and apart from becoming complete dropouts from our society, we must play the game. People lose faith in their religion because they cannot deal with the contradictions that result. This is particularly true of men, perhaps because of their more competitive natures. Only 5% of men in Europe attend church regularly. Men attend in somewhat higher numbers in the U.S., but they are still outnumbered by women. To many men concepts like "turning the other cheek" or "doing good to those who misuse you" are simply non-workable if not nonsensical. Doing good for others requires the promise of a tangible reward. The strong bond that exists between men who have served on the front lines during warfare is built upon something akin to love, and the men often refer to their "love" for their comrades. That sense of love and willingness to sacrifice and risk one's life for a comrade in arms is based upon reciprocity: the knowledge that others will sacrifice and perhaps die protecting you.

We should accept people who are different or who hold different beliefs or viewpoints, not because it will bring a reward in heaven but because it will reduce crime and violence. This benefits you and me directly. Bashing gay people because they are sinning runs counter to the goal of social harmony. Forget sin and act in a way that quells social unrest. That will benefit me and my family directly. Sacrifice for others with the expectation that they will sacrifice for you in a time of need. Go out of your way to be courteous on the highway because it improves your chances of surviving. Selfish cooperation works in all other animal societies, there is no reason it cannot work in ours if we will change our attitudes toward altruistic behavior and actively work toward the betterment of others for our own benefit.

What applies to individuals also applies to nations. Wealthy nations should share far more than they do with poor nations, not out of unconditional altruism, but because it is in their best interest to do so. Poverty is at the root of much of the unrest and violence in the world today. Helping to alleviate the causes of violence is a direct benefit to the people of wealthy nations. Enemy nations should seek every opportunity to reach out to each other and to understand each other rather than constantly making threats. The benefits to both sides are obvious.

Religion is sometimes seen as a refuge from life's problems. It should not be. Religion should be down in the dirt helping to solve the problems that beset us. Unfortunately today, organized religion is creating more enmity, divisiveness, social unrest, and violence than cooperation and harmony. In the United States fights over gay marriage, abortion, germ cell research, prayers and the teaching of religious doctrine in the public schools, and other peripheral issues are distracting us from the real issues which are tearing the world apart. We are playing Nero's fiddle while the world burns. On the international stage we are destroying the lives of thousands of people in an effort to protect needed resources or to force our political system and way of life on others. These are purely selfish motives and must be condemned as antisocial behavior.

We could achieve far more through selfish cooperation. We must stop categorizing behavior as sinful or blessed of God and start thinking in terms of cooperative social behavior which will promote harmony between individuals and nations. The only way to live up to the injunction to love our neighbor as ourselves is if the peoples of the world begin to realize, as the Athapaskan Indians realized, that we cannot stand alone.

The Axial sages did not create a new paradigm of social behavior. Cooperative behavior is as old as humanity itself. These standards of behavior are in each of us. The Axial sages correctly recognized that the changes wrought by the agricultural revolution lead humanity to emphasize aspects of its nature formerly kept under control by the need to act cooperatively in order to survive. Their solution was to develop rules of behavior which require us to transcend our biological nature through unconditional love in the hopes of achieving eternal peace and harmony. In this they failed completely. The social violence they sought to quell has never abated. Altruistic behavior has an underlying basis in selfishness through kinship and reciprocity. Until we are willing to recognize this very basic aspect of our nature and act accordingly we will never achieve the social harmony and cooperation which enabled our ancestors to survive and pass on the gift of life to us.

Coda

So where do we stand in this 200,000 year of our history as a species? Technologically we have risen to the level of gods, or so our ancient ancestors would believe if they could visit us, and we continue to move forward with a speed that almost defies our ability to assimilate. Scientifically we have advanced well beyond fearing routine natural phenomena and have begun to explore the stars. Socially, we have made some progress in adjusting from living in small, widely separated tribes to surviving in mega-cities with tens of millions of people.

Are we wealthier than the ancients? Yes, in material goods. Are we happier? I don't believe so. The little I know about people living a stone-age existence today in the Amazon or other remote areas of the world indicates that they are a particularly well fed, secure, and happy people who are able to feed themselves quite well on a few hours work each day. We, on the other hand, work ourselves to death, depriving family and friends of the quality time the ancients took for granted. Are we more secure? Perhaps, but our insecurities still overwhelm us. Do we enjoy a deeper, more enriching spiritual life? Almost certainly, no. The ancients did not compartmentalize the activities of their lives the way we do. They did not go to church, synagogue, mosque or temple one day a week and work the rest. Their religion was found in the things of the natural world, for which they had the utmost respect and reverence. They lived in a world of spirits and communed with them on an hourly basis. I envy them their faith; it had meaning for their lives in a comprehensive way that ours does not.

Have we adapted well to the extremely changed conditions which we find ourselves in as a result of agriculture and technology? In this I believe we still have a long way to go. Some aspects of the human nature that served our ancient ancestors so well are threatening to destroy us. We carry within us the tribal mentality that makes the members of "our" group people to be loved, protected, and cherished, while the members of "their" group are to be suspect, manipulated, and destroyed if it serves our purposes. We also carry within us the need to intimidate others through the use of power and the accumulation of material goods. And like the ancients we eat and eat and eat when food is available, always preferring the foods highest in calories, lest there be no food found tomorrow.

Most damaging of all, we continue our efforts to extract the maximum amount of resources from the environment in the shortest possible time with no thought for the future. Given the small size of the human population in ancient times and their primitive technologies, this did not appreciably harm their environment, or if it did, they could simply move on and let the damage repair itself. Today we are applying this approach on a world-wide scale with devastating results. We have not yet learned how to control those elements of our basic nature that enabled our ancestors to survive but which are maladaptive in modern society. We are stone-age creatures destroying the Garden of Eden with modern technology and toying with the most lethal weapons ever envisioned. Sooner or later we will probably use them against each other.

Science can in no way help us to outgrow or discipline those aspects of our nature which are no longer adaptive, only religion can do that if it chooses. I am not optimistic that it will so choose, for that would mean breaking from the barnacle-encrusted traditions and practices of the past, giving up the thirst for power and wealth, accepting women as full partners in the search for spiritual and social well being, and respecting and cherishing the spiritual search of others when it differs from our own. Only a revolution of the highest order could bring about such change, for people cling to power, wealth, position, and tradition with a tenacity that defies description.

The first step to improving the human condition is to accept without exception that we are biological creatures of this world, and that we have an evolutionary history. Locked in our genetic material is a remembrance of ancient battles won, of creative forces merging and clashing to propel life forward, sometimes at a snail's pace, at others with breath-taking speed. It's all there, waiting to be read – the whole story of life's struggle to survive – to beat the forces of destruction wrought by a hostile universe – a continuing drama more than 4 billion years old. No one knows how it will end for us as a species, but we, as the current masters of the planet, can influence the future course of this titanic endeavor – can continue to be a part of it if we are wise and choose our strategies carefully. We may fail, and if we do the drama will continue without us. That would be no tragedy. We will have had our day, as the dinosaurs had theirs, will have lived and suffered and triumphed in turn, and will have left our mark on the progression of earthly life.

Bibliography

Achebe, C. 1958. *Things Fall Apart*. New York: Anchor.

Allegretti, J. 2000. *The Plague Psalms*. Providence: Grim Reaper Books.

Armstrong, K. *The Great Transformation; The Beginning of our Religious Traditions.* New York: Alfred A. Knopf.

Behe, M.J. 1996. *Darwin's Black Box: The Biochemical Challenge to Evolution*. New York: The Free Press.

Buss, D. M. 2005. *The Strategies of Human Mating.* In: Sherman, P.W. and J. Alcock: *Exploring Animal Behavior: Readings From American Scientist.* 4th Ed. Pp.248-259.

Davis, P. and D.H. Kenyon. 1993. *Of Pandas and People: the Central Question of Biological Origins.* Richardson, Texas: Foundation for Thought and Ethics.

Davison, Lois A. 1954. *Ask of the Eagle*. Lewisburg: Bucknell University Press.

Dawkins, R. 1976. *The Selfish Gene*. Oxford: Oxford University Press.

Durant, W. 1953. The Renaissance. New York: Simon & Schuster.

Eiseley, L. 1975. *All the Strange Hours*. New York: Charles Scribner's Sons.

_____. 1971. *The Night Country*. New York: Charles Scribner's Sons.

Gilkey, L. 1993. Nature, Reality, and the Sacred. P181. Minneapolis: Fortress Press.

Howell, K.J. 2002. *God's Two Books*. Notre Dame: University of Notre Dame Press.

Johanson, D. and B. Edgar. 1996. *From Lucy to Language*. New York: Simon & Schuster Editions.

Kostelanetz, R. 1971. *Human Alternatives; Visions for us Now*. New York: Marrow.

Kozlovsky, D.G. 1974. *An Ecological and Evolutionary Ethic*. Englewood Cliffs: Prentice-Hall.

Manji, I. 2003. *The Trouble with Islam*. New York: St. Martin's Press.

Miller, G. 2000. *The Mating Mind*. New York: Anchor Books.

Napier, J. 1993. *Hands*. Princeton: Princeton University Press.

Niemark, P. 1993. *The Way of the Orisa*. New York: HarperCollins.

Pruitt, W. O. 1967. *Wild Harmony: Animals of the North*. New York: Nick Lyons Books.

Quinn, D. 1992. *Ishmael*. New York: Bantam Books.

Sears, P. B. 1937. *Deserts on the March*. Norman: University of Oklahoma Press.

Shermer, M. 1997. *Why People Believe Weird Things*. New York: MJF Books.

———. 2000. *How We Believe*. New York: Henry Holt and Company.

Townsend, R.F. and R.V. Sharp (eds.). 2004. New Haven and London: Yale University Press.

Ward, P.D. and D. Brownlee. 2000. *Rare Earth: Why Complex Life is Uncommon in the Universe*. New York: Springer-Verlag.

Wilson, E. O. 1968. *Consilience: The Unity of Knowledge*. New York: Alfred A. Knopf.

Wilson, F. R. 1998. *The Hand*. New York: Pantheon Books.

Zahavi, A. and A Vishag. 1997. *The Handicap Principle*. New York and Oxford: Oxford University Press.

CPSIA information can be obtained at www.ICGtesting.com
Printed in the USA
LVOW13s1054300813

350375LV00001B/99/P